どう言う？　こう解く！

新版 英語対訳で読む「算数・数学」入門

マイプラン 編
Gregory Patton 英文執筆

実業之日本社

PREFACE

The purpose of this book is to read the basic contents of "arithmetic and mathematics" we studied in elementary school and junior high school in simple English. We described as simply as possible so that even those who are not good at "arithmetic and mathematics" can read and understand smoothly. You can review both "arithmetic and mathematics" and English——You can kill two birds with one stone in this book.

In this book, we numbered each Japanese sentence and English sentence to make it easy to compare the 2 languages. We underlined the main English words and put the Japanese translations under them. Please refer to the translations when you read the English sentences.

Between Japanese and English, there are differences in how to read numbers and expressions. "Fractions" is a typical example. In Japanese, $\frac{2}{3}$ is read "sanbunnoni," from a denominator to a numerator in order. In English, it is read "2 over 3," from a numerator to a denominator in order. (cf. p.26-27) Generally, the word "times" is used in multiplication, and the word "divided by" is used in division. (cf. p.12-13) This book will be more fun when you read it with interest in the differences between Japanese and English in "arithmetic and mathematics."

Also, people in English speaking countries don't study exactly the same contents of "arithmetic and mathematics" as we study in Japan. For example, "Calculation of cranes and turtles" (cf. p.46-47) and "Probabilities (Drawing lots)" (cf. p.184-185) are the original contents in Japan. We hope you realize the strategy of Mr. Gregory Patton, who translated them into English.

Mori Satoshi

はじめに

本書は、小学校・中学校で学習した「算数・数学」の基礎的な内容を、平易な英語で読むための本です。「算数・数学」をニガテにしていた方でもスラスラと読み解けるように、できる限りやさしく記述しました。「算数・数学」のおさらいもでき、英語のおさらいもできる――そんな一石二鳥の一冊となっています。

本書では、英文、日本文にそれぞれ番号をつけて、英文と日本文を対照して読み比べられるようにしてあります。また、英文の主要な単語には下線を引いて、日本語の説明を加えました。英文を読むときの参考にしてください。

日本語と英語では、数字や数式の読み方に違いがあります。たとえば分数。日本語では、$\frac{2}{3}$を「３分の２」と分母、分子の順に読みますが、英語では 2 over 3 と、分子、分母の順に読みます(p.26、27)。一般的にかけ算は times、わり算は divided by を使います(p.12、13)。「算数・数学」の日本語と英語の表記の違いに興味をもって読んでもらえると、よりおもしろいと思います。

また、日本で学習する「算数・数学」とまったく同じ内容を、英語圏で学習しているわけではありません。「つるかめ算(p.46、47)」、「確率(くじ引き)(p.184、185)」などは日本独自の内容です。英訳をお願いした Gregory Patton 先生の苦心のあともお楽しみいただければと思います。

森　智史

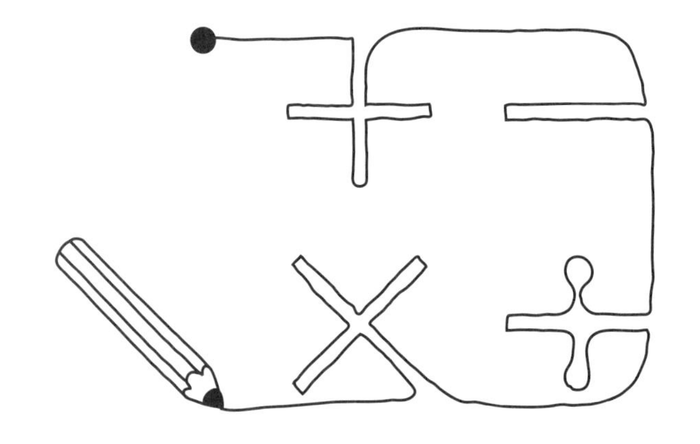

装幀／杉本欣右
イラスト／奥下和彦
ＤＴＰ／ユニックス＋サッシイ・ファム
本文執筆・編集／マイプラン＋森智史
校正／ Zen factory
編集協力／荻野守（オフィスＯＮ）

Contents
目　次

Contents

Chapter 1 Arithmetic
第1章 算 数

1. Four Arithmetic Operations ―― 12
四則計算
2. Large numbers ―― 14
大きい数
3. Round numbers ―― 16
概数
4. Categorizing whole numbers ―― 18
整数のなかま分け
5. Equal or greater than - Equal or less than - Less than ―― 20
以上・以下・未満
6. Decimals ―― 22
小数
7. Calculating Decimals ―― 24
小数の計算
8. Fractions ―― 26
分数
9. Addition and subtraction of fractions ―― 28
分数のたし算・ひき算
10. Multiplication and division of fractions ―― 30
分数のかけ算・わり算
11. Rates and percentages ―― 32
歩合と百分率
12. Problem of ratio ―― 34
割合の問題
13. Problem of proportion ―― 36
比の問題
14. Problem of the size of the per unit amount ―― 38
単位量あたりの大きさの問題
15. Speed ―― 40
速さ
16. Calculation of average ―― 42
平均算
17. Calculation of traveler's speeds, time, and distances ―― 44
旅人算
18. Calculation of cranes and turtles ―― 46
つるかめ算
19. Job problem ―― 48
仕事算
20. Age problem ―― 50
年齢算
21. Trees problem ―― 52
植木算

Contents

22. Calendar calculation ―― 54
日暦算
23. Formulas of figures ―― 56
図形の公式
24. Sizes of angles ―― 58
角の大きさ
25. Areas of triangles ―― 60
三角形の面積
26. Areas and circumferences of circles ―― 62
円の面積と円周
27. Numbers of cases ―― 64
場合の数
28. Combinations ―― 66
組み合わせ
29. Data dispersion and central value ―― 68
資料の散らばりと代表値

Column Is there the “kuku” also in English? ―― 70
「九九」は英語にもあるのか

Chapter 2 Numbers and expressions
第2章　数と式

30. Positive and negative numbers ―― 72
正負の数
31. Positive and negative numbers (question) ―― 74
正負の数(問題)
32. How to express expressions with letters ―― 76
文字式の表わし方
33. How to calculate expressions with letters ―― 78
文字式の計算
34. Use of expressions with letters ―― 80
文字式の利用
35. Prime Factorization ―― 82
素因数分解
36. Square roots ―― 84
平方根
37. Calculation of square roots ―― 86
√の計算
38. Multiplication and division of polynomials (expansion) ―― 88
多項式の乗除(展開)
39. Factorization ―― 90
因数分解
40. How to solve linear equations ―― 92
1次方程式の解き方

Contents

41. Linear equations word problem (too many or too few) — 94
1次方程式の文章題(過不足)

42. How to solve simultaneous equations (addition and subtraction method) — 96
連立方程式の解き方(加減法)

43. How to solve simultaneous equations (substitution method) — 98
連立方程式の解き方(代入法)

44. Simultaneous equations word problem (salt solutions) — 100
連立方程式の文章題(食塩水)

45. How to solve quadratic equations — 102
2次方程式の解き方

46. Quadratic equations and the solution formula — 104
2次方程式と解の公式

Column How to express large numbers in English — 106
英語では大きい数をどのように表わすのか

Chapter 3 Functions
第3章 関 数

47. Proportion and inverse proportion — 108
比例と反比例

48. Use of proportion (word problem) — 110
比例の利用(文章題)

49. Use of proportion (graph) — 112
比例の利用(グラフ)

50. Proportion and inverse proportion (graph) — 114
比例と反比例(グラフ)

51. Linear functions (introduction) — 116
1次関数(導入)

52. Problem of finding an equation of linear functions — 118
1次関数の式を求める問題

53. Problem of finding an intersection of the graphs of linear functions — 120
1次関数のグラフの交点を求める問題

54. Use of linear functions (problem of a moving point) — 122
1次関数の利用(動点の問題)

55. $y=ax^2$ (introduction) — 124
$y=ax^2$(導入)

56. $y=ax^2$ word problem (ratio of change) — 126
$y=ax^2$の文章題(変化の割合)

57. $y=ax^2$ word problem (problem of finding areas) — 128
$y=ax^2$の文章題(面積を求める問題)

Column Binary scale, decimal scale, hexadecimal scale, sexagesimal scale — 130
2進法, 10進法, 16進法, 60進法

Contents

Chapter 4 Figures
第4章 図 形

58. Basic vocabulary and concepts of figures —— 132
図形の基本語句と概念
59. Figure transformation —— 134
図形の移動
60. Parallel lines and angles —— 136
平行線と角
61. Parallel lines and areas —— 138
平行線と面積
62. Circles and sectors —— 140
円とおうぎ形
63. Inscribed angles —— 142
円周角
64. Tangent lines of circles —— 144
円の接線
65. Positional relationship of straight lines and planes —— 146
直線と平面の位置関係
66. Regular polyhedrons —— 148
正多面体
67. Projection drawing —— 150
投影図
68. Volume of conic solids —— 152
錐体の体積
69. Surface areas of cones —— 154
円錐の表面積
70. Solids of revolution —— 156
回転体
71. Congruent conditions of triangles —— 158
三角形の合同条件
72. Special parallelograms —— 160
特別な平行四辺形
73. Proof of congruence —— 162
合同の証明
74. Similarity conditions of triangles —— 164
三角形の相似条件
75. Proof of similarity —— 166
相似の証明
76. Midpoint theorem —— 168
中点連結定理
77. The Pythagorean theorem —— 170
三平方の定理
78. The Pythagorean theorem (Altitudes of regular triangles) —— 172
三平方の定理(正三角形の高さ)

Contents

79. The Pythagorean theorem (Diagonal lines of rectangular solids) —— 174
三平方の定理(直方体の対角線)

Column How to remember the circular constant —— 176
円周率の覚え方

Chapter 5 Use of data
第5章　データの活用

80. Frequency distribution table · Cumulative frequency —— 178
度数分布表・累積度数
81. Statistical probability —— 180
統計的確率
82. Probabilities (dice) —— 182
確率(さいころ)
83. Probabilities (Drawing lots) —— 184
確率(くじ引き)
84. Probability of not happening —— 186
起こらない確率
85. Box plots and interquartile range —— 188
箱ひげ図と四分位範囲
86. Sample surveys and complete surveys —— 190
標本調査と全数調査
87. Approximate values and significant figures —— 192
近似値と有効数字

Column The origin of commonly used letters in mathematics —— 194
数学でよく使われる文字の由来

Chapter 6 High school mathematics for beginners
第6章　高校数学入門

88. How to solve an inequality —— 196
不等式の解き方
89. Sets —— 198
集合
90. Quadratic functions —— 200
2次関数
91. Trigonometric ratios —— 202
三角比
92. Differential calculus —— 204
微分
93. Integral calculus —— 206
積分

X + Y = Z

Chapter 1
Arithmetic

第1章　算　数

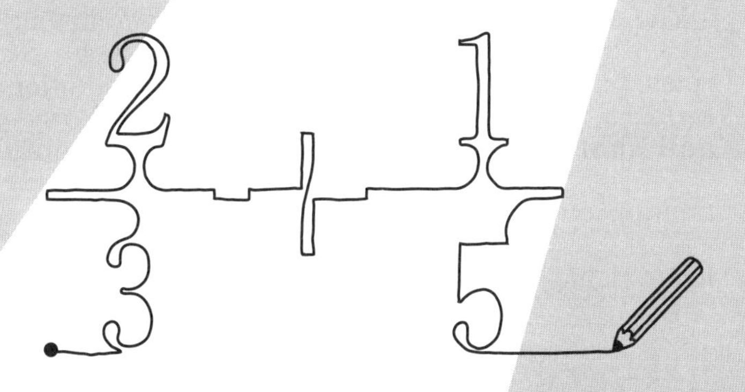

1. Four Arithmetic Operations

① Addition, subtraction, multiplication and division
たし算　ひき算　かけ算　わり算
calculations are called the four arithmetic operations.
計算　呼ばれる　四則計算
② "3 plus 4 equals 7" is a kind of addition.
3＋4＝7　～のひとつ
③ Subtraction is like "9 minus 2 equals 7."
～のような　9－2＝7
④ This is multiplication, "6 times 5 equals 30."
6×5＝30
⑤ "56 divided by 7 equals 8" is a kind of division.
56÷7＝8
⑥ In case we can't divide evenly by division, we
～の場合は　わる　均等に
write it like "25÷3=8···1," and the answer is read "8 with the remainder of 1."
あまり
⑦ The four arithmetic operations have the following
以下の
rules. ⑧ In expressions consisting of only addition
ルール　式　なり立っている
and subtraction, and ones consisting of only multiplication and division, we calculate them from
計算する　左から右へ
left to right.

① ② $2+5-3=7-3=4$
(2 plus 5 minus 3 equals 7 minus 3 equals 4)

① ② $2\times 6\div 3=12\div 3=4$
(2 times 6 divided by 3 equals 12 divided by 3 equals 4)

⑨ In expressions consisting of addition, subtraction, multiplication and division, we calculate the multiplication and division before (～より先に) the addition and subtraction.

$$\overset{\boxed{2}}{4+}\underline{15\overset{\boxed{1}}{\div}3}=4+5=9$$

(4 plus 15 divided by 3 equals 4 plus 5 equals 9)

⑩ In expressions with parentheses (かっこ), we first (先に) calculate the expressions within (～の中で) the parentheses.

$$12\overset{\boxed{2}}{\div}(\underline{4\overset{\boxed{1}}{+}2})=12\div6=2$$

※(12 divided by the quantity, 4 plus 2, close quantity, equals 12 divided by 6 equals 2)

1．四則計算

①たし算・ひき算・かけ算・わり算の計算を四則計算といいます。②「3＋4＝7」はたし算の1つです。③ひき算は「9－2＝7」のようなものです。④「6×5＝30」これはかけ算です。⑤「56÷7＝8」はわり算の1つです。⑥わり算でわり切れないときは、「25÷3＝8…1」のように表わし、「8あまり1」と読みます。⑦四則計算では以下のルールがあります。⑧たし算とひき算だけからなる式、またはかけ算とわり算だけからなる式では、左から右へ計算します。⑨たし算・ひき算・かけ算・わり算からなる式では、かけ算とわり算をたし算とひき算より先に計算します。⑩かっこがある式は、かっこの中の式を先に計算します。

※かっこがある式を英語では the quantity (かっこ), ～, close quantity (かっこ閉じる) と表わします。

● 「ケタ違い」な表現法、知ってました？

2. Large numbers

① The position to the left of ten million is called one hundred million.

位　1000万(10の100万)　呼ばれる　1億(100の100万)

② One hundred million is the number that ten million is gathered 10 times and is written, 100,000,000.

数　集められた　～回　書かれる

③ The number 247,861,950,000 is a combination of 2478 one hundred millions and 6195 ten thousands.

組み合わせ　1万

④ If we multiply one hundred million by 10, we get one billion.

かける　得る　10億

⑤ 10 times one billion is ten billion.

～倍　100億

⑥ 10 times ten billion is one hundred billion.

1000億

⑦ One hundred billion multiplied by 10 is one trillion and is written, 1,000,000,000,000.

かけられた　1兆

⑧ The number 5,121,926,400,000,000 is a combination of 5121 trillions and 9264 one hundred millions.

⑨ If we multiply one trillion by 10, we get ten trillion.

10兆

⑩ 10 times ten trillion is one hundred trillion.

100兆

⑪ 10 times one hundred trillion is one quadrillion.
1000兆

⑫ In this way, every time a position shifts to the left by 1 digit, a whole number becomes a multiple of 10.
このように　～するごとに　(位置が)変わる
桁　整数　～になる　倍数

2. 大きい数

①1000万の位の左の位を1億の位といいます。

②1億は1000万を10個集めた数で、100000000と書きます。

③247861950000は、1億を2478個と1万を6195個合わせた数です。

④1億を10倍した数は10億です。

⑤10億を10倍した数は100億です。

⑥100億を10倍した数は1000億です。

⑦1000億を10倍した数は1兆といい、1000000000000と書きます。

⑧5121926400000000は、1兆を5121個と1億を9264個合わせた数です。

⑨1兆を10倍した数は10兆です。

⑩10兆を10倍した数は100兆です。

⑪100兆を10倍した数は1000兆です。

⑫このように、整数は、位が1つ左に進むごとに、10倍になっています。

●「四捨五入」すると……、年齢ではやらないほうが？!

3. Round numbers

① Approximate numbers are called round numbers.
およその　数　呼ばれる　概数

② When we express approximate numbers, we use a
～するとき　表わす

method called "rounding off."
方法　四捨五入

③ In round numbers, the numbers,0, 1, 2, 3 and 4

are rounded down.
切り捨てられる

④ And the numbers, 5, 6, 7, 8 and 9 are rounded up.
切り上げられる

⑤ When we express the number 3124 as a round
～として

number of the first digit from the top, we round off
1つ目の　桁　上

the second digit from the top.
2つ目の

⑥ Because the second digit is 1, we round it down
～なので

and the number becomes 3000.
～になる

⑦ When rounding off the number 71829 to the
概数で表わす

second digit from the top, we round off the third
3つ目の

digit from the top.

⑧ Because the third digit is 8, we round it up and the

round number becomes 72000.

⑨ If a number rounded off to the second digit from the top becomes 4500, the whole number before it is rounded off is in the range 4450 to 4549.

If: ～なら
whole number: 整数
before: ～する前の
in the range 4450 to 4549: 4450と4549の間に

3. 概数

①およその数のことを概数といいます。
②概数で表わすとき、「四捨五入」という方法を使います。
③四捨五入では、数字が０、１、２、３、４のときは切り捨てます。
④そして、数字が５、６、７、８、９のときは切り上げます。
⑤3124を上から１桁の概数で表わすときは、上から２桁目の数字を四捨五入します。
⑥上から２桁目の数字は１だから、切り捨てて3000になります。
⑦71829を上から２桁の概数で表わすときは、上から３桁目の数字を四捨五入します。
⑧上から３桁目の数字は８だから、切り上げて72000になります。
⑨上から２桁の概数で表わした数が4500のときは、四捨五入する前のもとの整数は、4450と4549の間の数です。

● 「素数」って、わり切れないやつですね

4. Categorizing whole numbers

① Whole numbers can be divided into even numbers
整数　分けられる　～に　偶数
and odd numbers.
奇数

② Whole numbers such as 2, 4 and 6, which can be
～のような　わられる
divided by 2 are called even numbers.
～によって　呼ばれる

③ Whole numbers such as 1, 3 and 5, which leave a
残す
remainder of 1 when they are divided by 2 are called
あまり
odd numbers.

④ All whole numbers are either even numbers or
～か…のどちらか
odd numbers.

⑤ A whole number that can divide another whole
わる　別の
number is called a divisor.
約数

⑥ Because 8 can be divided by 1, 2, 4 and 8, these
～なので
numbers are divisors of 8.

⑦ The number that we get when we multiply a
得る　～するとき　かける
whole number some times is called a multiple of
何倍か　倍数
that number.

⑧ 16, 24 and 32 are the numbers when 8 is multiplied 2
かけられた
times, 3 times and 4 times, so they are multiples of 8.
～回
⑨ Also, whole numbers other than 1, which can only
また ～以外 だけ
be divided by 1 and itself are called prime numbers.
それ自身 素数
⑩ The number of divisors of prime numbers is 2.

⑪ 2, 3, 5, 7, 11 and so on are prime numbers and
～など
there are 25 prime numbers between 1 and 100.
～がある ～と…の間

4. 整数のなかま分け

①整数は、偶数と奇数に分けることができます。②2、4、6のように、2でわり切れる整数を偶数といいます。③1、3、5のように、2でわると1あまる整数を奇数といいます。④すべての整数は、偶数か奇数のどちらかになります。⑤ある整数をわることができる整数を、その数の約数といいます。⑥8は、1や2、4、8でわることができるので、これらの数は8の約数です。⑦ある整数を何倍かした整数を、その数の倍数といいます。⑧16や24、32は8を2倍、3倍、4倍した数なので、8の倍数です。⑨また、1以外の整数で、1とその数自身でしかわることのできない整数を素数といいます。⑩素数の約数の数は2個です。⑪2、3、5、7、11などが素数で、1から100までの整数の中に25個の素数があります。

●数学「未満」の話しですが……

5. Equal or greater than - Equal or less than - Less than

① When we show a range of numbers or compare
～するとき　表わす　範囲　数　比べる
the size of numbers, we use the terms "equal or
大きさ　言葉　等しい
greater than", "equal or less than" and "less than".
～より大きい　～より小さい　(単独で)～未満

② "Equal or greater than" shows that number and numbers larger than that number.
より大きい

③ For example, equal or greater than 8 is 8, 9, 10 and
たとえば　～など
so on and shows 8 and numbers larger than 8.

④ "Equal or less than" shows that number and numbers smaller than that number.
より小さい

⑤ For example, equal or less than 8 is 8, 7, 6 and so on and it shows 8 and numbers smaller than 8.

⑥ "Less than" shows numbers smaller than that number and does not include that number.
含む

⑦ For example, less than 8 is 7, 6, 5 and so on and means numbers less than 8.
意味する

⑧ The combination of the terms "equal or greater
組み合わせ

than", "equal or less than" and "less than" can show a range of numbers.

⑨ The whole numbers which are equal or greater than 5 and equal or less than 8 are 5, 6, 7 and 8.

整数

⑩ The whole numbers which are equal or greater than 5 and less than 8 are 5, 6 and 7.

5. 以上・以下・未満

①数の範囲を表わすときや、数の大きさを比べるときには、「以上」や「以下」、「未満」という言葉を使います。②「以上」は、その数と、その数よりも大きい数を表わします。③たとえば、8以上の数は、8、9、10などの数で、8と、8よりも大きい数になります。④「以下」は、その数と、その数よりも小さい数を表わします。⑤たとえば、8以下の数は、8、7、6などの数で、8と、8よりも小さい数を表わします。⑥「未満」は、その数よりも小さい数を表わし、その数は含みません。⑦たとえば、8未満の数は、7、6、5などの数で、8よりも小さい数になります。⑧「以上」や「以下」、「未満」を組み合わせると、数の範囲を表わすことができます。⑨5以上8以下の整数は、5、6、7、8です。⑩5以上8未満の整数は、5、6、7です。

●小数にも「位」があって……

6. Decimals

①The numbers, zero point four, thirty two point six eight, two point nine zero four and so on are called decimals. ②In decimals, "." is called a decimal point and the numbers to the right of the decimal point are read one by one without position. ③0.1 (zero point one) is one part of 1 divided into 10 equal parts. ④0.01 (zero point zero one) is one part of 1 divided into 100 equal parts. ⑤0.001 (zero point zero zero one) is one part of 1 divided into 1000 equal parts. ⑥The position directly to the right of the decimal point is called the first decimal place. ⑦The position to the right of the first decimal place is the second decimal place and the position to the right of the second decimal place is the third decimal place. ⑧The first decimal place is also called the tenths place, the second decimal place is called the hundredths place, and the third

numbers 数 / zero point four 0.4 / thirty two point six eight 32.68 / two point nine zero four 2.904 / and so on 〜など / are called 呼ばれる / decimals 小数 / decimal point 小数点 / are read 読まれる / one by one 1つずつ / without 〜なしで / position 位 / part 部分 / divided 分けられた / into 〜に / equal 等しい / directly すぐ / first decimal place 小数第一位 / second decimal place 小数第二位 / third decimal place 小数第三位 / also 〜もまた / tenths place $\frac{1}{10}$の位 / hundredths place $\frac{1}{100}$の位

decimal place is called the thousandths place.
$\frac{1}{1000}$の位

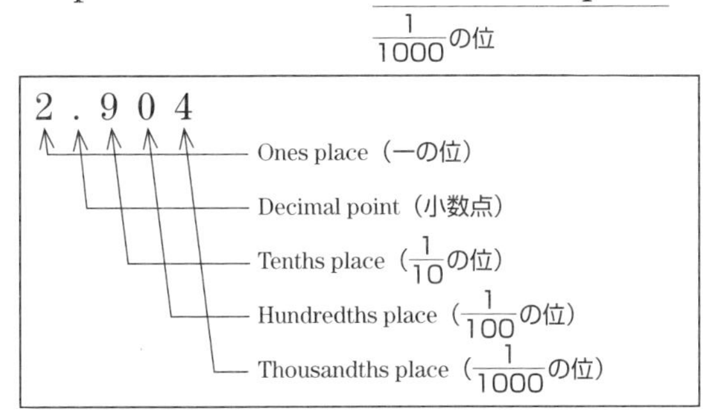

⑨In the same way as whole numbers, multiplying decimals by 10 moves the position up one digit and multiplying by 1 over 10 moves the position down one digit.

（In the same way as：～と同じように　multiplying：かけること　moves：上げる　digit：桁　1 over 10：$\frac{1}{10}$　moves：下げる）

6．小数

①0.4や32.68、2.904などの数を小数といいます。②小数の「.」を小数点といい、小数点より右にある数は、位をつけずに1つずつ読みます。③0.1は、1を10等分した1つ分です。④0.01は、1を100等分した1つ分です。⑤0.001は、1を1000等分した1つ分です。⑥小数点のすぐ右にある位を小数第一位といいます。⑦小数第一位の右の位を小数第二位、小数第二位の右の位を小数第三位といいます。⑧小数第一位は$\frac{1}{10}$の位、小数第二位は$\frac{1}{100}$の位、小数第三位は$\frac{1}{1000}$の位ともいいます。⑨小数も整数と同じように、10倍すると位が1桁上がり、$\frac{1}{10}$にすると、位が1桁下がります。

● 「桁」をそろえることが肝心

7. Calculating Decimals

① In addition of decimals and subtraction of decimals,
たし算　小数　ひき算
we align the decimal point and add or subtract
そろえる　小数点　たす　ひく
numbers in the same position. ② When the right end
位　端
decimal position of the answer is 0, it is omitted.
省かれる

25.81	(twenty five point eight one)
+ 3.39	(plus three point three nine)
29.20	(twenty nine point two)

10.32	(ten point three two)
− 4.24	(minus four point two four)
6.08	(six point zero eight)

③ In multiplication of decimals, first, we ignore the
かけ算　まず　無視する
decimal points and calculate in the same way as
計算する　～と同じように
whole numbers. ④ After that, we find the sum of the
整数　和
number of decimal places in the multiplier and in
かける数
the multiplicand, and insert a decimal point the same
かけられる数　挿入する　同じ
number of place into the answer of the calculation.

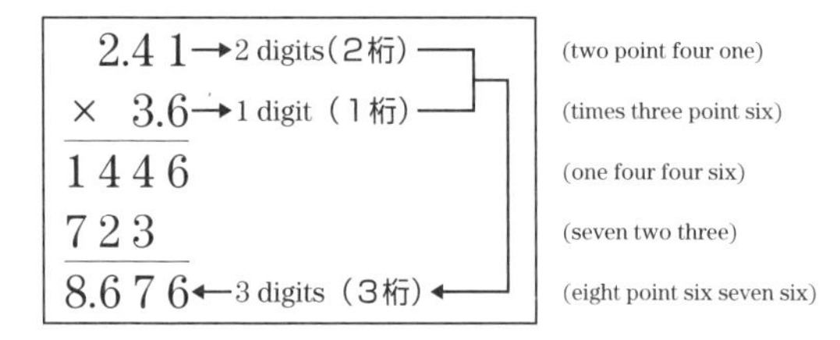

⑤ In division of decimals, in order to make the divisor a whole number, we move the decimal point to the right, and then, in the dividend as many decimal places as in the divisor and calculate the answer. ⑥ We insert a decimal point into the answer, aligning with the position of the decimal point in the dividend.

division わり算　in order to ～するために　make the ～にする　divisor わる数　dividend わられる数　as many ～と同じ数だけ　aligning そろえながら

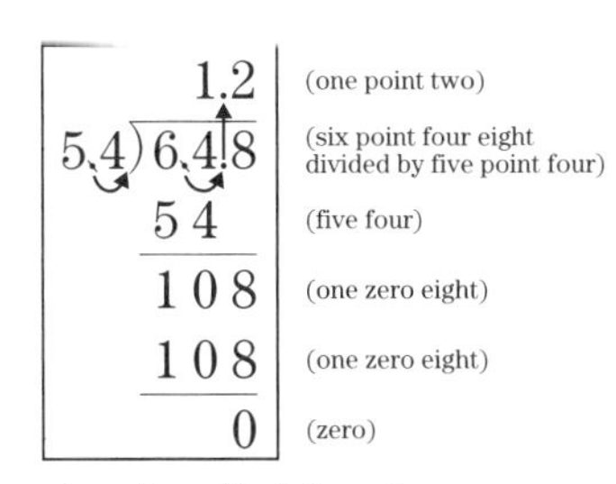

7．小数の計算

①小数のたし算やひき算は、小数点の位置をそろえて、同じ位の数をたしたりひいたりします。②答えの小数部分の右端が0のときは、0を省(はぶ)きます。③小数のかけ算は、まず、小数点がないものとして、整数と同じように計算します。④そのあと、小数点以下の桁数が、かける数とかけられる数の小数点以下の桁数の和になるように、その計算の答えに小数点を打ちます。⑤小数のわり算は、わる数が整数になるように、わる数の小数点以下の桁数だけ、わられる数の小数点を右に移して計算します。⑥わられる数の移した小数点の位置にそろえて、計算の答えに小数点を打ちます。

8. Fractions

① 1 over 3, 2 over 3, 2 over 5 and so on are called fractions.

$\frac{1}{3}$　～など　呼ばれる　分数

② In fractions with the same denominator, the larger the numerator becomes, the larger the number becomes.

同じ　分母　より大きい　分子　～になる　数

$$\frac{2}{5}<\frac{3}{5}$$

(2 over 5 is less than 3 over 5)

③ In fractions with the same numerator, when the denominator becomes smaller, the number becomes larger.

～するとき　より小さく

$$\frac{1}{3}>\frac{1}{5}$$

(1 over 3 is greater than 1 over 5)

④ Fractions like 2 over 3 and 4 over 5 which are less than 1 are called proper fractions. ⑤ In proper fractions, the denominator is larger than the numerator. ⑥ Fractions like 3 over 3 or 8 over 5 which are equal to 1 or greater than 1 are called improper fractions. ⑦ Improper fractions can be expressed as the sum of whole numbers and proper fractions. ⑧ Fractions

～のような　～より少ない　真分数　等しい　より大きい　仮分数　表わされる　和　整数

$$\frac{17}{6}=2\frac{5}{6}$$

(17 over 6 equals 2 and 5 over 6)

which are expressed as the sum of whole numbers
～として
and proper fractions are called mixed fractions.
帯分数
⑨ In fractions, even when both the denominator and
～と…の両方
numerator are multiplied by the same number other
かけられる ～以外
than 0, the size doesn't change. ⑩ Also, when both
大きさ 変わる また
the denominator and the numerator are divided by
わられる
the same number other than 0, the size doesn't change.

8．分数

①$\frac{1}{3}$や$\frac{2}{3}$、$\frac{2}{5}$などの数を分数といいます。②分母が同じ分数は、分子が大きいほうが、数が大きくなります。③分子が同じ分数は、分母が小さいほうが、数が大きくなります。④$\frac{2}{3}$や$\frac{4}{5}$のように、1より小さい分数を真（しん）分数といいます。⑤真分数は、分子よりも分母の数のほうが大きくなります。⑥$\frac{3}{3}$や$\frac{8}{5}$のように、1と等しいか、1よりも大きい分数を仮（か）分数といいます。⑦仮分数は、整数と真分数の和で表わすことができます。⑧整数と真分数の和で表わした分数は、帯（たい）分数といいます。⑨分数は、分母と分子に0以外の同じ数をかけても、大きさは変わりません。⑩また、分母と分子を0以外の同じ数でわっても大きさは変わりません。

9. Addition and subtraction of fractions

① In addition of fractions with the same denominator, the denominator is left as it is, and the numerators are added together.

たし算 (addition)　分数 (fractions)　同じ (same)　分母 (denominator)　残される (is left)　そのまま (as it is)　分子 (numerators)　たされる (are added)　いっしょに (together)

$$\frac{1}{5}+\frac{3}{5}=\frac{4}{5}$$

(1 over 5 plus 3 over 5 equals 4 over 5)

② In subtraction of fractions with the same denominator, the denominator is left as it is, and the numerators are subtracted.

ひき算 (subtraction)　ひかれる (are subtracted)

$$\frac{6}{7}-\frac{2}{7}=\frac{4}{7}$$

(6 over 7 minus 2 over 7 equals 4 over 7)

③ In addition and subtraction of fractions with different denominators, we calculate after making the denominators the same.

異なった (different)　計算する (calculate)　～を…にする (making)

$$\frac{1}{3}+\frac{2}{5}=\frac{5}{15}+\frac{6}{15}=\frac{11}{15}$$

(1 over 3 plus 2 over 5 equals 5 over 15 plus 6 over 15 equals 11 over 15)

④ In addition and subtraction of mixed fractions, there are two ways.

帯分数 (mixed fractions)　～がある (there are)　方法 (ways)

⑤ One way is to separately calculate the whole numbers and proper fractions.
分けて　整数　真分数

$$3\frac{1}{4}+1\frac{2}{3}=3\frac{3}{12}+1\frac{8}{12}=4\frac{11}{12}$$

(3 and 1 over 4 plus 1 and 2 over 3 equals 3 and 3 over 12 plus 1 and 8 over 12 equals 4 and 11 over 12)

⑥ Another way is to change the mixed fractions into the improper fractions and calculate.
もう1つの　変える　～に　仮分数

$$2\frac{1}{3}-1\frac{2}{5}=\frac{7}{3}-\frac{7}{5}=\frac{35}{15}-\frac{21}{15}=\frac{14}{15}$$

(2 and 1 over 3 minus 1 and 2 over 5 equals 7 over 3 minus 7 over 5 equals 35 over 15 minus 21 over 15 equals 14 over 15)

9．分数のたし算・ひき算

①分母が同じ分数のたし算は、分母はそのままで、分子どうしをたします。②分母が同じ分数のひき算は、分母はそのままで、分子どうしをひきます。③分母が異なる分数のたし算やひき算は、分母を同じにしてから計算をします。④帯分数のたし算やひき算の計算では、２通りの方法があります。⑤１つは、整数と真分数に分けて計算する方法です。⑥もう１つは、帯分数を仮分数になおしてから計算する方法です。

● 「約分」するって、どういうこと？

10. Multiplication and division of fractions

① In multiplying fractions, we multiply each numerator together and each denominator together.

かけるとき　分数　それぞれの　分子　いっしょに　分母

$$\frac{2}{3}\times\frac{4}{5}=\frac{2\times4}{3\times5}=\frac{8}{15}$$

(2 over 3 times 4 over 5 equals 2 times 4 over 3 times 5 equals 8 over 15)

② In whole numbers, we calculate as a fraction with a denominator of 1.

整数　計算する　～として

$$2\times\frac{3}{7}=\frac{2}{1}\times\frac{3}{7}=\frac{2\times3}{1\times7}=\frac{6}{7}$$

(2 times 3 over 7 equals 2 over 1 times 3 over 7 equals 2 times 3 over 1 times 7 equals 6 over 7)

③ In mixed fractions, we calculate after converting them into an improper fraction.

帯分数　変える　～に　仮分数

$$1\frac{2}{3}\times\frac{2}{7}=\frac{5}{3}\times\frac{2}{7}=\frac{5\times2}{3\times7}=\frac{10}{21}$$

(1 and 2 over 3 times 2 over 7 equals 5 over 3 times 2 over 7 equals 5 times 2 over 3 times 7 equals 10 over 21)

④ When we can reduce in the middle of a calculation, we calculate after reducing.

～するとき　約分する　～の途中で　計算

$$\frac{2}{3}\times\frac{5}{6}=\frac{\overset{1}{\cancel{2}}\times5}{3\times\underset{3}{\cancel{6}}}=\frac{5}{9}$$

(2 over 3 times 5 over 6 equals 2 times 5 over 3 times 6 equals 5 over 9)

⑤ When the product of two numbers is 1, one is called the reciprocal of the other.

積　数　呼ばれる　逆数　もう一方

⑥The reciprocal of a fraction is a fraction whose denominator and numerator are switched.
入れかえられた (are switched)

⑦The reciprocal of 5 over 3 becomes 3 over 5.
$\frac{5}{3}$ (5 over 3)　$\frac{3}{5}$ (3 over 5)

⑧In dividing fractions, we calculate by multiplying the dividend by the reciprocal of the divisor.
わるとき (In dividing)　わられる数 (dividend)　わる数 (divisor)

$$\frac{1}{6}\div\frac{5}{3}=\frac{1}{6}\times\frac{3}{5}=\frac{1\times\overset{1}{\cancel{3}}}{\underset{2}{\cancel{6}}\times5}=\frac{1}{10}$$

(1 over 6 divided by 5 over 3 equals 1 over 6 times 3 over 5 equals 1 times 3 over 6 times 5 equals 1 over 10)

10. 分数のかけ算・わり算

①分数のかけ算は、分子どうし、分母どうしをそれぞれかけます。
②整数は、分母が1の分数として計算します。
③帯分数は、仮分数になおしてから計算します。
④計算の途中で約分ができるときは、約分をしてから計算します。
⑤2つの数の積が1となるとき、一方の数を、もう一方の数の逆数といいます。
⑥分数の逆数は、分母と分子を入れかえた分数になります。
⑦$\frac{5}{3}$の逆数は、$\frac{3}{5}$になります。
⑧分数のわり算は、わられる数にわる数の逆数をかけて計算します。

●イチローの「打率」、英語で言えますか？

11. Rates and percentages

① When we compare two amounts, the number that
～するとき 比べる 量 数
expresses how many times the amount is, compared
表わす 何倍 比べられた
to the original amount is called the ratio.
もとの 呼ばれる 割合

② Ratio can be expressed not only with whole
表わされる ～だけでなく 整数
numbers, decimals, and fractions but also with
小数 分数 ～も
buai and percentages.
歩合(英語にない) 百分率

③ *Buai* is the way to express 1 over 10 of the original
方法 $\frac{1}{10}$
amount as 1 *wari*, 1 over 100 as 1 *bu*, and 1 over
～として 割 分
1000 as 1 *rin*.
厘

④ When expressed in *buai*, ratios which are expressed as decimals and fractions, zero point six
0.6
becomes 6 *wari*, zero point four five becomes 4
0.45
wari 5 *bu* and 1 over 8 becomes 1 *wari* 2 *bu* 5 *rin*.

⑤ Percentages are the way to express the ratio of the original amount treated as 100, and zero point
扱われた
zero one is 1%.

⑥ When the ratio is expressed as a percentage, the ratio expressed as a decimal or a fraction is multiplied (かけられる) by 100.

⑦ From zero point seven times 100 equals 70 ($0.7 \times 100 = 70$), zero point seven becomes 70%, and from 1 over 4 times 100 equals 25, 1 over 4 becomes 25%.

11. 歩合と百分率

①2つの量を比べるとき、比べる量がもとにする量の何倍かを表わした数を割合といいます。②割合は、整数や小数、分数以外に、歩合や百分率で表わすことができます。
③歩合は、もとにする量の$\frac{1}{10}$を1割、$\frac{1}{100}$を1分、$\frac{1}{1000}$を1厘とする表わし方です。④小数や分数で表わされた割合を歩合で表わすと、0.6は6割、0.45は4割5分、$\frac{1}{8}$は1割2分5厘となります。
⑤百分率は、もとにする量を100としたときの割合の表わし方で、0.01を1%とします。⑥割合を百分率で表わすときは、小数や分数で表わされた割合を100倍します。⑦0.7は、$0.7 \times 100 = 70$から70%、$\frac{1}{4}$は、$\frac{1}{4} \times 100 = 25$から25%となります。

●どこまでをいくつに「分ける」かが問題

12. Problem of ratio

Question

① There is a book with 120 pages.
～がある　120ページある本

② On the first day 2 over 5 of the whole book were read, and on the second day 3 over 8 of the remainder were read.
1日目　$\frac{2}{5}$　全体の　読まれた　2日目　残り

③ At this time, how many unread pages are left?
このとき　いくつの　読まれていない　残されている

How to solve

④ Because the number of pages read on the first day is 2 over 5 of 120 pages, it is (120 times 2 over 5 equals) 48 pages.
～なので　数　$\left(120\times\frac{2}{5}=\right)48$

⑤ In this case, there are (120 minus 48 equals) 72 pages not yet read.
(120−48=)72　まだ～ない

⑥ Because the number of pages read on the second day is 3 over 8 of 72 pages, it is (72 times 3 over 8 equals) 27 pages.

⑦ So, the number of pages not yet read is (72 minus 27 equals) 45 pages.

Answer ⑧ 45 pages

Caution

⑨ Thinking the number of pages read on the second day is 3 over 8 of 120 pages, let's be careful not to do like this,(120 times 3 over 8 equals) 45 pages.

Thinking：考えて　careful：注意深い　like this：〜のように

12. 割合の問題

問 題 ①120ページの本があります。②1日目に全体の$\frac{2}{5}$を読み、2日目に残りの$\frac{3}{8}$を読みました。③このとき、読んでいないページは、あと何ページありますか。

解き方 ④1日目に読んだページ数は、120ページの$\frac{2}{5}$だから、$120\times\frac{2}{5}=48$(ページ)です。⑤このとき、まだ読んでいないページは、$120-48=72$(ページ)あります。⑥2日目に読んだページ数は、72ページの$\frac{3}{8}$だから、$72\times\frac{3}{8}=27$(ページ)です。⑦よって、まだ読んでいないページは、$72-27=45$(ページ)となります。

答 え ⑧45ページ

注 意 ⑨2日目に読んだページ数を、120ページの$\frac{3}{8}$と考えてしまい、$120\times\frac{3}{8}=45$(ページ)としないように注意しましょう。

●いつも他人と「比べて」ばかりいるけれど……

13. Problem of proportion

Question

① The proportion of the money that Yoshio and
比
Kazuya have was 4 to 3.
4：3

② Because Yoshio bought a book, the proportion of
～なので　買った
the money Yoshio and Kazuya have became 8 to 7
and the sum of the money the two have became
合計
9000 yen.
円

③ How many yen is the price of the book Yoshio
何円　値段
bought?

How to solve

④ Because the money Yoshio has after buying the
～のあと
book is 8 over 15 of the total, it is (9000 times 8 over
$\frac{8}{15}$　全体　$\left(9000\times\frac{8}{15}=\right)4800$
15 equals) 4800 yen.

⑤ When we express the money Yoshio had before
～するとき　表わす　～の前
buying the book as blank yen, using the proportion,
～として　□
4 to 3 equals blank to 4200 (9000 minus 4800).
4：3=□：4200(9000−4800)

⑥ Because 4200 divided by 3 equals 1400, the money Yoshio had before buying the book is (1400 times 4 equals) 5600 yen.

4200÷3=1400

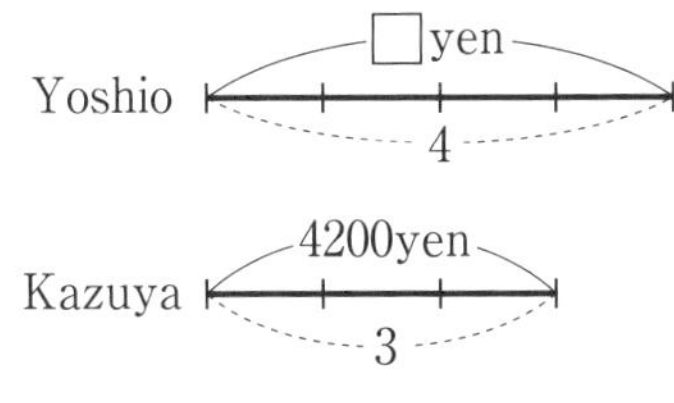

⑦ The price of the book Yoshio bought becomes (5600 minus 4800 equals) 800 yen.

Answer ⑧ 800 yen

13．比の問題

問 題 ①よしおさんとかずやさんの所持金の比は４：３でした。②よしおさんは本を買ったので、よしおさんとかずやさんの所持金の比は８：７となり、２人の所持金の合計は9000円になりました。③よしおさんが買った本の値段は何円ですか。

解き方 ④よしおさんが本を買ったあとの所持金は、全体の$\frac{8}{15}$だから、$9000\times\frac{8}{15}=4800$(円)です。⑤よしおさんが本を買う前の所持金を□円として比を使って表わすと、４：３＝□：4200(9000－4800)となります。⑥4200÷3＝1400だから、よしおさんが本を買う前の所持金は、1400×4＝5600(円)です。⑦よしおさんが買った本の値段は、5600－4800＝800(円)となります。

答 え ⑧800円

14. Problem of the size of the per unit amount

Question

① There is a car which can drive 600 kilometers per 40 liters.
～がある　キロメートル(km)　～あたり　リットル(L)

② At a gas station it costs 2400 yen to put 15 liters of gasoline into the car.
ガソリンスタンド　(費用が)かかる　入れる　ガソリン　～に

③ How much yen's worth of gasoline is needed for this car to drive 360 kilometers when gasoline is put into the car at this gas station?
何円分の～　必要とされる　入れられる

How to solve

④ Let's think about the distance the car can drive per 1 liter of gasoline and the cost of 1 liter of gasoline.
道のり　値段

⑤ This car can drive (600 divided by 40 equals) 15 kilometers per 1 liter of gasoline.
(600÷40=)15

⑥ In addition, the cost of 1 liter of gasoline at this gas station is (2400 divided by 15 equals) 160 yen.
さらに

⑦ For this car to drive 360 kilometers, (360 divided
この自動車が360キロメートル走るためには

by 15 equals) 24 liters of gasoline is needed.

⑧ So, to drive 360 kilometers, (160 times 24 equals) 3840 yen's worth of gasoline is needed.
(160×24=)3840

Answer ⑨3840 yen

14. 単位量あたりの大きさの問題

問 題 ①40L で600km 走る自動車があります。②あるガソリンスタンドで、この自動車にガソリン15L を入れると2400円かかります。③このガソリンスタンドでこの自動車にガソリンを入れるとき、この自動車で360km 走るためには何円分のガソリンを入れる必要がありますか。

解き方 ④ガソリン1L あたりで走ることのできる道のりと、ガソリン1L あたりの値段を考えます。⑤この自動車は、ガソリン1L あたりで、600÷40＝15 (km)走ることができます。⑥また、このガソリンスタンドでのガソリン1L あたりの値段は、2400÷15＝160 (円)です。⑦この自動車で360km 走るためには、360÷15＝24 (L)のガソリンが必要になります。⑧よって、360km 走るためには、160×24＝3840 (円)分のガソリンが必要となります。

答 え ⑨3840円

●速さとキョリと時間の親密な「三角関係」とは？

15. Speed

① Speed can be found by dividing the distance
速さ　見つけられる　わること　道のり
travelled by the time taken. ② The units to express
進んだ　かかった　単位　表わす
speed are speed per hour, speed per minute, speed
時速　分速　秒速
per second and so on. ③ Speed per hour is the speed
～など
expressing the distance travelled in 1 hour. ④ From
1時間で
120 divided by 3 equals 40, the speed of a vehicle
120÷3=40　車
traveling a distance of 120 kilometers in 3 hours is
キロメートル(km)
40 kilometers per hour. ⑤ Speed per minute is the
時速40キロメートル
speed expressing the distance travelled in 1 minute.

⑥ From 800 divided by 5 equals 160, the speed of a
800÷5=160
bicycle traveling a distance of 800 meters in 5
自転車　メートル(m)
minutes is 160 meters per minute. ⑦ Speed per
分速160メートル
second is the speed expressing the distance travelled in 1 second. ⑧ From 200 divided by 25 equals 8, the speed of a person running 200 meters
人間
in 25 seconds is 8 meters per second.
秒速8メートル

⑨When the speed is known, from the speed, the
〜するとき　知られている
distance travelled and the time taken can be found.

⑩A vehicle traveling 30 kilometers per hour will travel (30 times 3 equals) 90 kilometers in 3 hours.
(30×3=)90

⑪At speed of 150 meters per minute, a distance of 900 meters takes (900 divided by 150 equals) 6 minutes.
(900÷150=)6

15. 速さ

①速さは、進んだ道のりをかかった時間でわると求めることができます。②速さを表わす単位には、時速や分速、秒速などがあります。③時速は、1時間あたりに進む道のりで表わした速さです。④120kmの道のりを3時間で走る車の速さは、120÷3=40より、時速40kmです。⑤分速は、1分間あたりに進む道のりで表わした速さです。⑥800mの道のりを5分で走る自転車の速さは、800÷5=160より、分速160mです。⑦秒速は、1秒間あたりに進む道のりで表わした速さです。⑧200mを25秒で走る人の速さは、200÷25=8より、秒速8mです。⑨速さがわかっているときは、速さから、進んだ道のりやかかる時間を求めることができます。⑩時速30kmで走る車は、3時間で、30×3=90(km)進みます。⑪900mの道のりを分速150mで走ると、900÷150=6(分)かかります。

● 「平均点」はまあまあでした？

16. Calculation of average

Question

① Mamoru took tests of four subjects, Japanese,
受ける 教科 国語(日本語)
arithmetic, science and social studies.
算数 理科 社会
② As a result, the average score of the 4 subjects of
その結果 平均の 得点
Mamoru's Japanese, arithmetic, science and social studies, was 83 points.
点
③ In addition, the average score of the three
さらに
subjects Japanese, arithmetic and science was 81 points.

④ How many points was the score of Mamoru's
何点
social studies test?

How to solve

⑤ Because the average score of Mamoru's four
〜なので
subjects is 83 points, the sum of the scores of the 4
合計
subjects is (83 times 4 equals) 332 points.
(83×4=)332
⑥ Also, because the average score of the three
また

subjects, Japanese, arithmetic and science is 81 points, the sum of the scores of Japanese, arithmetic and science is (81 times 3 equals) 243 points.

⑦So, the score of the social studies test will be (332 minus 243 equals) 89 points.
(332−243=)89

Answer ⑧89 points

16. 平均算

問 題 ①まもるさんは、国語、算数、理科、社会の４教科のテストを受けました。②その結果、まもるさんの国語、算数、理科、社会の４教科の平均点は83点でした。③また、国語、算数、理科の３教科の平均点は81点でした。④まもるさんの社会のテストの点数は何点ですか。

解き方 ⑤まもるさんの４教科のテストの平均点は83点だから、４教科の点数の合計は、83×4＝332(点)です。⑥また、国語、算数、理科の３教科のテストの平均点は81点だから、国語、算数、理科の点数の合計は、81×3＝243(点)です。⑦よって、社会のテストの点数は、332－243＝89(点)となります。

答 え ⑧89点

●かめでもウサギに「追いつく」には？

17. Calculation of traveler's speeds, time, and distances

Question

① A little brother walked at a speed of 70 meters per
弟　メートル(m)　～につき
minute from his house to the station. ② 8 minutes
分
after the little brother left home, his older brother
～した後　出発した　兄
chased after him by bicycle at a speed of 210 meters
追いかけた　自転車　(two hundred ten)
per minute. ③ How many minutes after the little
何分
brother left home does the older brother catch up
～に追いつく
with his little brother?

How to solve

④ Before the older brother leaves home, the little
～するまで
brother travels (70 times 8 equals) 560 meters.
進む　(70×8=)560
⑤ Because the older brother travels at 210 meters
～なので
per minute and the little brother at 70 meters per
minute, the distance between them shortened in
～の間の　縮められる
one minute is (210 minus 70 equals) 140 meters.
(210−70=)140
⑥ The older brother catches up with the little brother,

(560 divided by 140 equals) 4 minutes after the older
(560÷140=)4
brother leaves home. ⑦Because the older brother leaves home 8 minutes after the little brother leaves home, the older brother catches up with the little brother, (4 plus 8 equals) 12 minutes after the little
(4+8=)12
brother leaves home.

Answer ⑧12 minutes after

17. 旅人算

問　題　①弟は、毎分70mの速さで歩いて、家から駅に向かいました。②兄は、弟が家を出発してから8分後に家を出発して、毎分210mの速さで、自転車で弟を追いかけました。③兄が弟に追いつくのは、弟が家を出発してから何分後ですか。

(解き方) ④兄が家を出発するまでに、弟は、70×8=560(m)進んでいます。⑤兄は毎分210m、弟は毎分70mの速さで進むので、1分間で縮まる2人の距離は、210−70=140(m)です。⑥兄が弟に追いつくのは、兄が家を出発してから、560÷140=4(分後)です。⑦兄が家を出発したのは、弟が家を出発してから8分後なので、兄が弟に追いつくのは、弟が家を出発してから、4+8=12(分後)です。

答　え ⑧12分後

●つるとかめが「計算」ですべった?!

18. Calculation of cranes and turtles

Question

① There is a total of 9 cranes and turtles.
合計 つる かめ

② The total number of legs is 26.
数 足

③ How many of each, cranes and turtles, are there?
何びき それぞれ ～がいますか

How to solve

④ Assuming all 9 are cranes, the number of legs is
推定すると

(2 times 9 equals) 18.
(2×9=)18

⑤ Actually because the number of legs is 26, the
実際には ～なので

difference between the number of legs assumed
～と…の間の差

in case all are cranes and the actual number of legs
～の場合に 実際の

is (26 minus 18 equals) 8 legs.
(26−18=)8

⑥ The difference between the number of one crane's legs and one turtle's legs is (4 minus 2 equals) 2.

⑦ Because the difference from the actual number is 8,
～との差

the number of turtles are (8 divided by 2 equals) 4.
(8÷2=)4

⑧ The number of cranes is (9 minus 4 equals) 5.

Answer ⑨ 5 cranes, 4 turtles

18. つるかめ算

問 題 ①つるとかめが合わせて９ひきいます。
②足の本数は合わせて26本です。
③つるとかめは、それぞれ何びきいますか。

解き方 ④９ひきすべてがつると仮定すると、足の本数は、2×9=18(本)です。
⑤実際には、足の本数は26本だから、すべてがつると仮定した場合との差は、26−18=8(本)です。
⑥つるとかめの１ぴきあたりの足の本数の差は、4−2=2(本)です。
⑦実際との足の本数の差が８本だから、かめは、8÷2=4(ひき)です。
⑧つるは、9−4=5(ひき)です。

答 え ⑨つる５ひき(羽)、かめ４ひき

●2人の「仕事量」がバレちゃった?!

19. Job problem

Question

① There is a job that takes 20 days to do alone by
～がある 仕事 かかる 1人で
Tatsuya or 30 days to do alone by Kazuya. ② On this job, at first Tatsuya works by himself for 5 days and
最初に 彼自身で
after that Tatsuya and Kazuya work together and
その後 いっしょに
finish the job. ③ All together, how many days does
終える 全部で 何日
it take to complete the job?
終える

How to solve

④ Tatsuya finishes 1 over 20 of the whole job in 1 day.
$\frac{1}{20}$ 全体の 1日で
⑤ If Tatsuya works for 5 days, he finishes (1 over 20
～なら $\left(\frac{1}{20}\times5=\right)\frac{1}{4}$
times 5 equals) 1 over 4 of the job. ⑥ Kazuya finishes 1 over 30 of the whole job in 1 day. ⑦ If Tatsuya and Kazuya work together, they finish (1 over 20 plus 1 over 30 equals 3 over 60 plus 2 over 60 equals 5 over
$\left(\frac{1}{20}+\frac{1}{30}=\frac{3}{60}+\frac{2}{60}=\frac{5}{60}=\right)\frac{1}{12}$
60 equals) 1 over 12 of the whole job. ⑧ The job left

over is (1 minus 1 over 4 equals) 3 over 4 of the job,
残されている
therefore the number of days they work together is
よって 数
(3 over 4 divided by 1 over 12 equals) 9 days.

⑨ Therefore, the number of days it takes to finish the job is, (5 plus 9 equals) 14 days.

Answer ⑩ 14 days

19. 仕事算

問 題 ①たつやさんが1人ですると20日かかり、かずやさんが1人ですると30日かかる仕事があります。②この仕事を、最初はたつやさんが1人で5日間働き、その後、たつやさんとかずやさんの2人で働いて仕事を終えます。③この仕事を終えるのに、全部で何日かかりますか。

解き方 ④たつやさんは、1日で全体の$\frac{1}{20}$を終わらせます。⑤たつやさんが5日間働くと、全体の$\frac{1}{20}\times5=\frac{1}{4}$が終わります。⑥かずやさんは、1日で全体の$\frac{1}{30}$を終わらせます。⑦たつやさんとかずやさんが2人で働くと、1日で全体の$\frac{1}{20}+\frac{1}{30}=\frac{3}{60}+\frac{2}{60}=\frac{5}{60}=\frac{1}{12}$を終わらせます。⑧残っている仕事は、$1-\frac{1}{4}=\frac{3}{4}$だから、2人で働いたときにかかる日数は、$\frac{3}{4}\div\frac{1}{12}=9$(日)です。⑨よって、この仕事を終えるのにかかる日数は、$5+9=14$(日)です。

答 え ⑩14日

●いつまでたっても「兄」は兄、「弟」は弟だけど……

20. Age problem

Question

① Now, Akira is 11 years old and Akira's father is 45 years old.
(Now: 現在)

② From now, how many years after will Akira's father's age be 3 times Akira's age?
(From now: 今から / how many years after: 何年後 / age: 年齢 / times: ～倍)

How to solve

③ The difference between the age of Akira's father and Akira is always (45 minus 11 equals) 34 years.
(difference: 差 / between … and: ～と…の間の / (45 minus 11 equals) 34: (45－11＝)34)

④ The difference in the two's age will be 2 times Akira's age when Akira's father's age becomes 3 times Akira's age.
(becomes: ～になる)

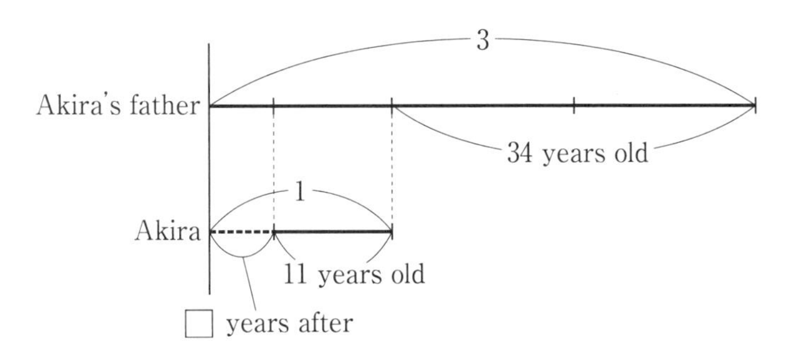

⑤ According to the Figure, when Akira's father's age is 3 times Akira's age, Akira is (34 divided by 2 equals) 17 years old.

(According to 〜によると / Figure 図 / (34 divided by 2 equals) 17 (34÷2=)17)

⑥ Now, Akira is 11 years old, so (17 minus 11 equals) 6 years later.

((17 minus 11 equals) 6 (17−11=)6)

Answer ⑦ 6 years later

20. 年齢算

問 題 ①現在、あきらさんは11歳、あきらさんのお父さんは45歳です。②あきらさんのお父さんの年齢が、あきらさんの年齢の3倍になるのは、今から何年後ですか。

(解き方) ③あきらさんのお父さんとあきらさんの年齢の差は、常に、45−11＝34(歳)です。④あきらさんのお父さんの年齢があきらさんの年齢の3倍になるとき、あきらさんの年齢の2倍が、2人の年齢の差になります。⑤図より、あきらさんのお父さんの年齢が、あきらさんの年齢の3倍になるとき、あきらさんの年齢は、34÷2＝17(歳)です。⑥現在、あきらさんは11歳だから、17−11＝6(年後)です。

答 え ⑦6年後

●池のまわりに「木」を植えたのは誰？

21. Trees problem

Question

① There is a straight road of 240 meters in length.
～がある まっすぐな 道 メートル(m) 長さ

② On one side of this road from end to end, cherry
片側 端から端まで サクラの

trees are planted every 15 meters.
木 植えられる ～おきに

③ In this case, how many cherry trees are needed?
この場合 いくつの 必要とされる

How to solve

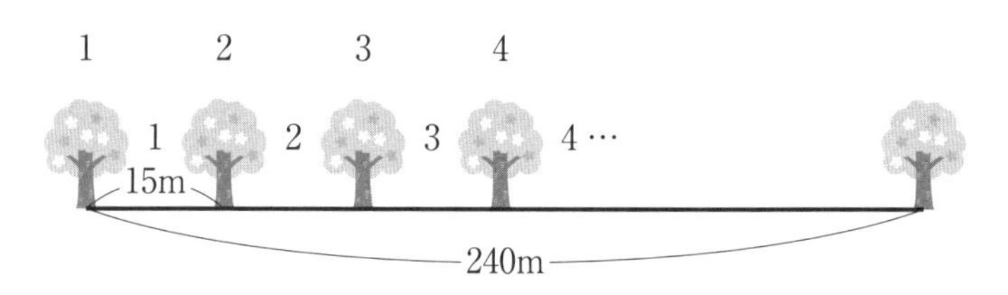

④ Because a cherry tree is planted at both ends of
～なので 両方の

the road, the number of cherry trees becomes the
数

number of spaces between the trees plus 1.
間隔 ～の間の 1本増し

⑤ When the cherry trees are planted, the number of
～するとき

spaces between the trees is (240 divided 15 equals) 16.
(240÷15=)16

⑥ Therefore the cherry trees needed are (16 plus 1
よって (16+1=)17

equals) 17 trees.

Answer ⑦ 17 trees

Caution

⑧ When there is a tree at both ends, the number of trees becomes 1 more than the number of spaces.
(more than: ～より多い)
⑨ Around a pond and the like, when the perimeter is connected, the number of trees and the number of spaces are equal.
(Around: ～のまわり / pond: 池 / the like: 同様のもの / perimeter: 周囲 / is connected: つなげられている / equal: 等しい)

21．植木算

問 題 ①長さが240mのまっすぐな道路があります。②この道路の片側に、15mおきにサクラの木を端から端まで植えます。③このとき、サクラの木は何本必要ですか。

解き方 ④道路の両端にもサクラの木を植えるので、サクラの木の本数は、木と木の間(あいだ)の数＋1となります。⑤サクラの木を植えたときの、木と木の間の数は、240÷15＝16です。⑥よって、サクラの木は、16＋1＝17(本)必要です。

答 え ⑦17本

注 意 ⑧両端に木があるときは、木の本数は間の数よりも1多くなります。⑨池のまわりなど、周囲がつながっているときは、木の本数と間の数が等しくなります。

●うるう「年」はメンドーだよね……

22. Calendar calculation

Question

① It is Wednesday, March 5 of a certain year.
3月 ある

② What day of the week is August 22 of this year?
何曜日 8月

How to solve

③ Because March has 31 days, including March 5,
～なので 含めると
there are (31minus 5 plus 1 equals) 27 days left in
～がある (31−5+1=)27 残されている
March.

④ There are 30 days in April, 31 days in May, 30 days
4月 5月
in June, and 31 days in July.
6月 7月

⑤ From March 5 to August 22, there are (27 plus 30 plus 31 plus 30 plus 31 plus 22 equals) 171 days.

⑥ 1 week has 7 days.

⑦ From 171 divided by 7 equals 24 with the remainder
171÷7=24…3
of 3, 171 days is 24 weeks and 3 days.

⑧ If we think the days from Wednesday to the next
～なら
Tuesday as one set, August 22 becomes Friday,
ひとまとまりとして

three days after Tuesday.
　　　　　　～後

Answer ⑨Friday

	月 MON	火 TUE	水 WED	木 THU	金 FRI	土 SAT	日 SUN
3	…	…	…	…	…	1	2
	3	4	5	6	7	8	9
	10	11	12	13	14	15	16
	17	18	19	20	21	22	23
	24	25	26	27	28	29	30
	31	…	…	…	…	…	…
4	…	1	2	3	4	5	6
	7	8	9	10	11	12	13
	14	15	16	17	18	19	20
	21	22	23	24	25	26	27
	28	29	30	…	…	…	…
	…	…	…	…	…	…	…
5	…	…	…	1	2	3	4
	5	6	7	8	9	10	11
	12	13	14	15	16	17	18
	19	20	21	22	23	24	25
	26	27	28	29	30	31	…
	…	…	…	…	…	…	…

	月 MON	火 TUE	水 WED	木 THU	金 FRI	土 SAT	日 SUN
6	…	…	…	…	…	…	1
	2	3	4	5	6	7	8
	9	10	11	12	13	14	15
	16	17	18	19	20	21	22
	23	24	25	26	27	28	29
	30	…	…	…	…	…	…
7	…	1	2	3	4	5	6
	7	8	9	10	11	12	13
	14	15	16	17	18	19	20
	21	22	23	24	25	26	27
	28	29	30	31	…	…	…
	…	…	…	…	…	…	…
8	…	…	…	…	1	2	3
	4	5	6	7	8	9	10
	11	12	13	14	15	16	17
	18	19	20	21	22	23	24
	25	26	27	28	29	30	31
	…	…	…	…	…	…	…

22. 日暦算

問　題　①ある年の3月5日は水曜日です。②この年の8月22日は何曜日ですか。

(解き方)　③3月は31日あるので、3月5日を含めると、3月は残り、31－5＋1＝27(日)あります。④4月は30日、5月は31日、6月は30日、7月は31日あります。⑤3月5日から8月22日までは、27＋30＋31＋30＋31＋22＝171(日)あります。⑥1週間は7日です。⑦171÷7＝24…3 より、171日は、24週間と3日です。⑧水曜日から次の火曜日までをひとまとまりと考えると、8月22日は、火曜日の3日後の金曜日となります。

答　え　⑨金曜日

●「台形」の面積の求め方、覚えてました？

23. Formulas of figures

① There are formulas to find areas and volume of figures.
～があります　公式　求める　面積　体積　図形

② (1) The area of a square is side times side.
正方形　1辺×1辺

③ (2) The area of a rectangle is height times width.
長方形　縦　横

④ (3) The area of a parallelogram is base times altitude.
平行四辺形　底辺　高さ

⑤ (4) The area of a rhombus is diagonal line times diagonal line divided by 2.
ひし形　対角線×対角線÷2

⑥ (5) The area of a trapezoid is the quantity upper base plus lower base, close quantity, times altitude divided by 2.
台形　（上底＋下底）

⑦ (6) The area of a triangle is base times altitude divided by 2.
三角形

⑧ (7) The volume of a cube is side times side times side. ⑨ (8) The volume of a cuboid is height times width times altitude.
立方体　直方体

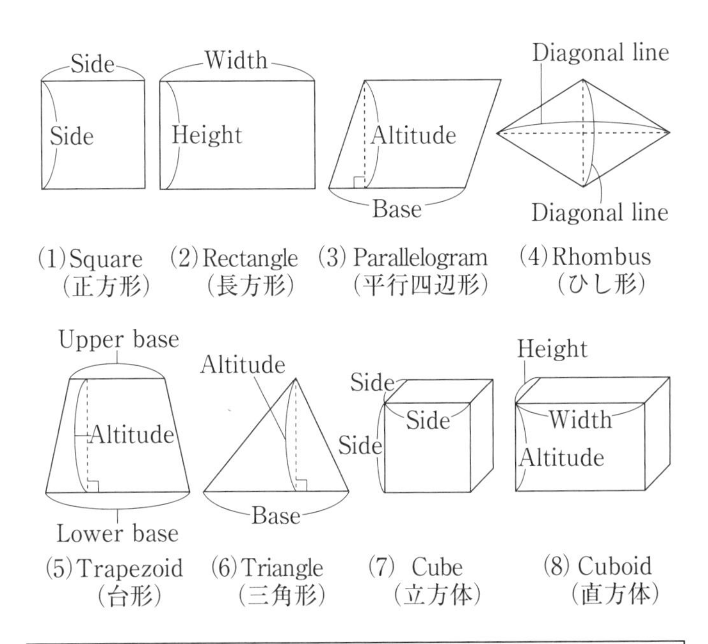

23. 図形の公式

①図形には、面積や体積を求めるための公式があります。
②(1) 正方形の面積　1辺×1辺
③(2) 長方形の面積　縦×横
④(3) 平行四辺形の面積　底辺×高さ
⑤(4) ひし形の面積　対角線×対角線÷2
⑥(5) 台形の面積　(上底＋下底)×高さ÷2
⑦(6) 三角形の面積　底辺×高さ÷2
⑧(7) 立方体の体積　1辺×1辺×1辺
⑨(8) 直方体の体積　縦×横×高さ

●三角形の「内角」の和は180度だから……

24. Sizes of angles

Question

① Find the sum of the sizes of the internal angles of an octagon.
求めなさい 和 大きさ 内角 八角形

How to solve

② Like triangles, quadrilaterals, pentagons …, a figure enclosed by only straight lines is called a polygon.
~のように 三角形 四角形 五角形 図形 囲まれた 直線 呼ばれる 多角形

③ The sum of the sizes of the internal angles of a polygon can be found in the case of *n*-gon by calculating 180 degrees times the quantity *n* minus 2, close quantity.
求められる 場合 n角形 計算すること 180°×(n−2)

④ This can be explained, that the quantity *n* minus 2, close quantity triangles can be made when a polygon is divided into triangles.
説明される つくられる ~するとき 分けられる ~に

Figure 1
(図1)

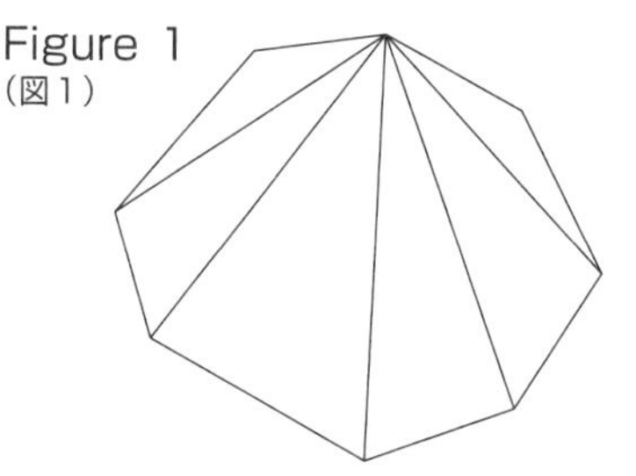

⑤ Like in Figure 1, an

octagon can be divided into 6 triangles.

⑥ So, the sum of the sizes of the internal angles of an octagon is (180 degrees times the quantity 8 minus 2, close quantity, equals) 1080 degrees.
(180°×(8−2)=)1080°

Answer

⑦ 1080 degrees

24. 角の大きさ

問 題 ①八角形の内角の大きさの和を求めなさい。

(解き方) ②三角形、四角形、五角形……のように、直線だけで囲まれた図形を多角形といいます。
③多角形の内角の大きさの和は、n 角形の場合、180°×(n−2)で求められます。
④これは、多角形を三角形に分けると、(n−2)個の三角形ができることから説明できます。
⑤図1のように、八角形では、6個の三角形ができます。
⑥よって、八角形の内角の大きさの和は、180°×(8−2)=1080°です。

答 え ⑦1080°

●三角形の面積の出し方には「高さ」が重要？

25. Areas of triangles

Question

① Find how many square centimeters the area of triangle ABC is.

求めなさい　何平方センチメートル(cm²)　面積　三角形

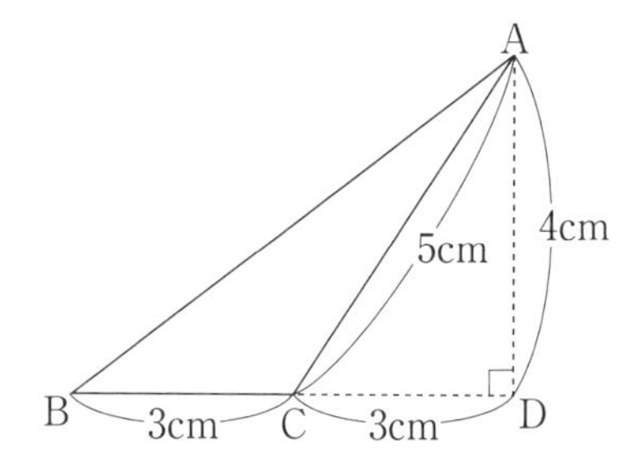

How to solve

② The area of a triangle can be found by calculating the base times the altitude divided by 2.

求められる　計算すること　底辺×高さ÷2

③ If we look at BC as the base, the altitude of triangle ABC is AD.

～なら　～として

④ So, the area is (3 times 4 divided by 2 equals) 6 square centimeters.

(3×4÷2=)6

⑤ Like in triangle ABC, the altitude of a figure is sometimes taken outside the figure.

～のように　ときどき　とられる　～の外側

⑥ Be careful not to mistake the altitude of triangle ABC as the 5 centimeters length of side AC.

注意する / ～しないように / まちがえる / 長さ / 辺

Answer

⑦ 6 square centimeters

25. 三角形の面積

問 題 ①三角形 ABC の面積は何 cm^2か、求めなさい。

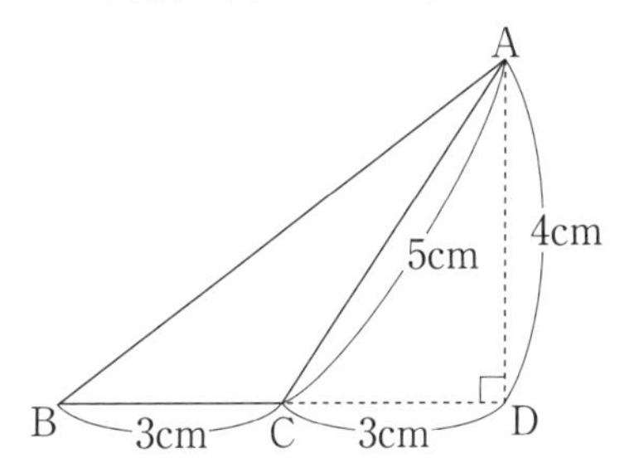

解き方 ②三角形の面積は、底辺×高さ÷2で求められます。
③三角形 ABC は、底辺を BC とみると、高さにあたるのは AD です。
④よって、面積は、3×4÷2＝6(cm^2)です。
⑤三角形 ABC のように、図形の高さは、その図形外にとることもあります。
⑥辺 AC の長さである 5 cm を三角形 ABC の高さとまちがえないように注意しましょう。

答 え ⑦ 6 cm^2

● 「円周率」を何桁まで覚えてますか？

26. Areas and circumferences of circles

Question

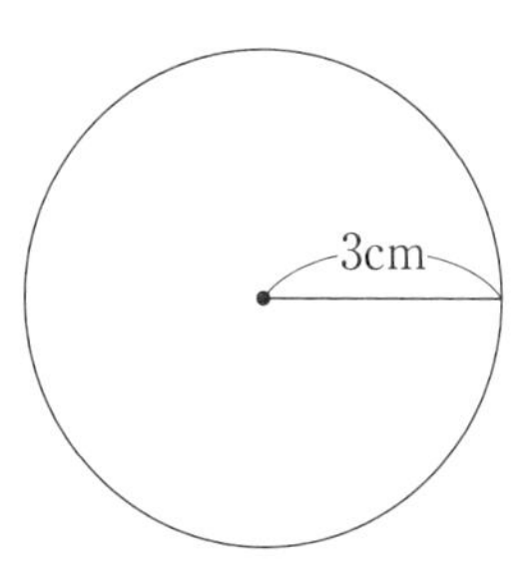

① How many centimeters is the length of the circumference of a circle with a radius of 3 centimeters?

何センチメートル(cm)　長さ　円周　円　半径

② Also, how many square centimeters is the area? ③ Find them respectively.

また　平方センチメートル(cm^2)　面積　求めなさい　それぞれ

④ The circular constant is recognized as 3.14.

円周率　理解される　～として

How to solve

⑤ The length of the circumference of a circle can be found by calculating the diameter times the circular constant.

求められる　計算すること　直径×円周率

⑥ The diameter of a circle with a radius of 3 centimeters is (3 times 2 equals) 6 centimeters.

(3×2=)6

⑦ So, the length of the wanted circumference is (6 times 3.14 equals) 18.84 centimeters.

求められている

⑧The area of a circle can be found by calculating the radius times the radius times the circular constant.

⑨So, the wanted area is (3 times 3 times 3.14 equals) 28.26 square centimeters.

⑩Be careful to use the diameter when finding the circumference and the radius when finding the area.

Be careful：注意する　when：～するとき

Answer

⑪circumference length：18.84 centimeters, area：28.26 square centimeters

26. 円の面積と円周

問　題　①半径が3cmである円の、円周の長さは何cmですか。②また、面積は何cm^2ですか。③それぞれ求めなさい。④円周率は、3.14とします。

解き方　⑤円の円周の長さは、直径×円周率で求められます。⑥半径が3cmである円の直径は、3×2＝6(cm)です。⑦よって、求める円周の長さは、6×3.14＝18.84(cm)です。⑧円の面積は、半径×半径×円周率で求められます。⑨よって、求める面積は、3×3×3.14＝28.26(cm^2)です。⑩円周を求めるときには直径を、面積を求めるときには半径を使うということに注意しましょう。

答　え　⑪円周の長さ：18.84cm、面積：28.26cm^2

●つい見落としてしまう「パターン」が……？

27. Numbers of cases

Question

① 4 persons, A, B, C and D line up side by side.
並ぶ　(横に)並んで

② How many ways in all, are there to line up?
何通り　全部で　～がありますか

How to solve

Figure 1 (図1)

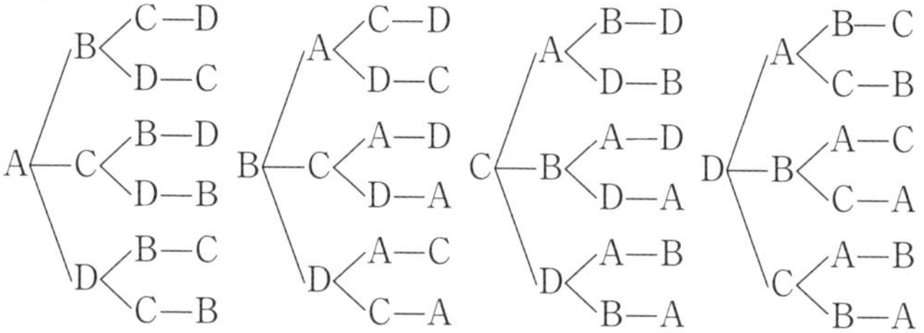

- A
  - B
    - C—D
    - D—C
  - C
    - B—D
    - D—B
  - D
    - B—C
    - C—B
- B
  - A
    - C—D
    - D—C
  - C
    - A—D
    - D—A
  - D
    - A—C
    - C—A
- C
  - A
    - B—D
    - D—B
  - B
    - A—D
    - D—A
  - D
    - A—B
    - B—A
- D
  - A
    - B—C
    - C—B
  - B
    - A—C
    - C—A
  - C
    - A—B
    - B—A

③ To examine every thinkable case, it is good thinking
調べる　すべての　考えられる　場合
to organize in good order by writing a tree diagram
整理する　順序　書くこと　樹形図
like in Figure 1. ④ From Figure 1, there are 24 ways
～のような　図
in all. ⑤ Also, we can think about it like the following
また　次の～
calculation. ⑥ Because 1 out of the 4 from A to D
計算　～なので　～のうち
comes to the farthest left side, there are 4 ways.
一番(遠い)　端
⑦ Coming to the second from the left, except for the
2番目　～を除いて

1 person lined up in the farthest left side, there are 3 ways. ⑧In the same way, third from the left becomes 2 ways and on the farthest right side becomes 1 way. ⑨So, in all cases, it is (4 times 3 times 2 times 1 equals) 24 ways.

同様に　3番目　(4×3×2×1=)24

Answer ⑩24 ways

27. 場合の数

問 題 ①A、B、C、Dの4人が横一列に並びます。②並び方は、全部で何通りありますか。

解き方 ③考えられるすべての場合を調べるには、図1のような樹形図を書いて、順序よく整理して考えるとよいです。④図1より、全部で24通りです。⑤また、次のように計算で考えることもできます。⑥一列のうち、一番左端にくるのは、AからDの4人のうち1人なので、4通りです。⑦左から二番目にくるのは、一番左端に並んだ1人を除いて3通りです。⑧同様に、左から三番目は2通り、一番右端は1通りとなります。⑨よって、すべての場合は、4×3×2×1=24(通り)です。

答 え ⑩24通り

●どの「組」になるかで、明暗が分かれるよね

28. Combinations

Question

① We decide leaders by choosing 2 persons from 5 persons, A, B, C, D and E.

decide 決める　leaders 係　choosing 選ぶこと

② How many ways in all can leaders be decided?

How many ways 何通り　in all 全部で　be decided 決められる

How to solve

③ In a case that 2 persons, A and B, become leaders and in a case that 2 persons, B and A, become leaders, if the combination is the same, we think of it as 1 way.

case 場合　if ～なら　combination 組み合わせ　same 同じ　as ～として

④ When we write a tree diagram that doesn't count duplicates, it looks like Figure 1.

When ～するとき　tree diagram 樹形図　count 数える　duplicates 重複するもの　like ～のように　Figure 図

Figure 1（図1）

A—B, A—C, A—D, A—E　B—C, B—D, B—E　C—D, C—E　D—E

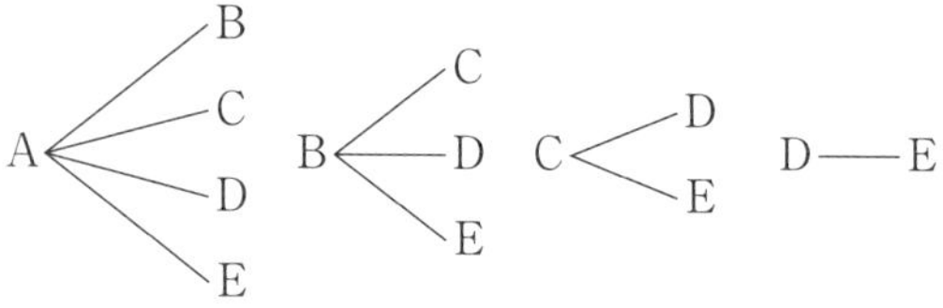

⑤ So, it is 10 ways in all.

⑥Also, we can think about it like the following calculation. ⑦When we choose 2 out of 5, for example if we count by distinguishing (A, B) from (B, A), it is (5 times 4 equals) 20 ways in all. ⑧Because the same combination is counted 2 times, the number of cases wanted is (20 divided by 2 equals) 10 ways.

また ～のように 次の～ 計算 ～中 ～と…を区別すること (5×4=)20 ～なので 数えられる ～回 数 求められる (20÷2=)10

Answer ⑨10 ways

28. 組み合わせ

問 題 ①A、B、C、D、Eの5人から、2人選んで係を決めます。②係は全部で何通りできますか。

解き方 ③AとBの2人が係になった場合と、BとAの2人が係になった場合など、組み合わせが同じ場合は、1通りと考えます。④重複するものを数えないように樹形図を書くと、図1のようになります。⑤よって、全部で10通りです。⑥また、次のように計算で考えることもできます。⑦5人中2人選ぶとき、たとえば(A, B)と(B, A)を区別して数えると、全部で5×4=20(通り)です。⑧これは、同じ組み合わせのものを2回数えることになるので、求める場合の数は、20÷2=10(通り)です。

答 え ⑨10通り

●あの「偏差値」も実はここから……？

29. Data dispersion and central value

Question

①The following table is a frequency distribution table of the height of 24 children in an elementary school. ②From this frequency distribution table, find (1) the category the median is included in, and (2) the mode.

次の～ 表 度数分布表 高さ 求めなさい 階級 中央値 含まれている 最頻値

Category (Height) (cm)	Frequency (Number of children)
Not less than Less than 150.0～155.0	3
155.0～160.0	4
160.0～165.0	9
165.0～170.0	6
170.0～175.0	2
Sum	24

How to solve

③A frequency distribution table is a table of frequency of each category when data of a certain group is divided into several categories. ④The median is the value of the middle when the data is arranged in order of sizes. ⑤Out of the data of the 24, the twelfth and thirteenth together are included

それぞれの ～するときの 資料 ある 集団 分けられる ～に いくつかの 真ん中の 並べられた 順番に 大きさ ～のうち 12番目 13番目 ともに

in the not less than 160 centimeters to less than 165
~以上 センチメートル(cm) ~未満
centimeters category. ⑥The mode refers to the middle
表わす
values of the category with the most frequency.
最も多い

Answer ⑦(1) the not less than 160 centimeters to less than 165 centimeters category (2) 162 point 5 centimeters

29. 資料の散らばりと代表値

問 題 ①次の表は、ある小学校の児童24人の身長の度数分布表です。②この度数分布表から、(1)中央値(メジアン)が含まれる階級、(2)最頻値(モード)をそれぞれ求めなさい。

階級(身長)(cm)	度数(人)
以上 未満 150.0~155.0	3
155.0~160.0	4
160.0~165.0	9
165.0~170.0	6
170.0~175.0	2
計	24

(解き方) ③度数分布表とは、ある集団の資料をいくつかの階級に分け、各階級の度数を表にしたものです。④中央値とは、資料の値を大きさの順に並べたときの中央の値のことです。⑤24個の資料のうち、中央の値となる12番目と13番目はともに160.0cm 以上165.0cm 未満の階級に含まれています。⑥最頻値は、度数の最も多い階級の真ん中の値です。

答 え ⑦(1)160.0cm 以上165.0cm 未満の階級
(2)162.5cm

Is there the "kuku" also in English?

「九九」は英語にもあるのか

① In Japan it is normal to memorize the "*kuku*" when in the second grade of elementary school. ② "*niningashi*, *nisangaroku*, *nishigahachi* ...," we memorize it in the way the words sound good. ③ Is there the "*kuku*" also in English? ④ In English speaking countries, they don't devise how to say like the "*kuku*" of Japan to memorize the multiplication. ⑤ Like "two times two is four, two times three is six, two times four is eight ...," they memorize by reciting as it is. ⑥ When you hear that, you might think, "Does multiplication take a lot of time for English speaking people?" ⑦ In fact, however, by continuing drilling, the answer "56" will flash instantly just by looking at the type face "7 times 8." ⑧ While Japanese are reciting "*shichihagojuuroku*," a person who learned multiplication in English will know the answer quickly.

---

①日本ではふつう、小学２年生のときに「九九」を覚えます。②「ににんがし、にさんがろく、にしがはち…」と、語呂のよい言い方にして覚えます。③英語にも「九九」はあるのでしょうか。④英語圏の国では、日本の「九九」のように、言い方を工夫してかけ算を暗記することはありません。⑤「two times two is four、two times three is six、two times four is eight…」のように、そのまま唱えて覚えるのです。⑥そのように聞くと、「英語圏の人はかけ算に時間がかかるのではないだろうか」と思うかもしれません。⑦しかし実際には、訓練を積むことにより、「7×8」といった字面を見ただけで「56」という答えが瞬時にひらめくようになります。⑧日本人が「しちはごじゅうろく」と唱えている間に、英語でかけ算をきちんと身につけた人なら、さっと答えがわかるのです。

Chapter 2
Numbers and expressions

第2章　数と式

●0より小さい数を「負の数」といいました……ね

30. Positive and negative numbers

① Numbers greater than 0 are positive numbers and
数　～より大きい　正の数
numbers less than 0 are negative numbers.
～より小さい　負の数

② 0 is neither a positive nor a negative number.
～でも…でもない

③ Negative numbers are expressed with a "−" and
表わされる　～をつけて
positive numbers are sometimes expressed with a "+".

④ On a number line expressing the size of numbers,
数直線　表わしている　大きさ
the right direction from the starting point is the
方向　原点
positive numbers and the left direction is the negative numbers.

⑤ On a number line, the numbers to the right become larger and the numbers to the left become
～になる　より大きく
smaller.
より小さく

⑥ The distance between a number and the starting
距離　～と…の間の
point is called that number's absolute value.
呼ばれる　絶対値

⑦ The absolute value of positive 6 is 6 and the
+6

absolute value of negative 3 is 3.
（negative 3：−3）

⑧ The absolute value of 0 is 0.

⑨ In positive numbers, the greater the absolute value is, the greater the number is, and in negative numbers, the greater the absolute value is, the smaller the number is.
（the greater ... ：大きくなればなるほど／the greater ... ：ますます大きく）

30．正負の数

①0より大きい数を正の数、0より小さい数を負の数といいます。②0は正でも負でもない数です。
③負の数には「−」をつけて表わし、正の数には「+」をつけて表わすことがあります。
④数の大きさを表わす数直線では、原点より右の方向が正の数、左の方向が負の数です。⑤数直線上では、右にある数ほど大きく、左にある数ほど小さくなります。
⑥数直線上で、ある数と原点との距離を、その数の絶対値といいます。⑦+6の絶対値は6で、−3の絶対値は3です。⑧0の絶対値は0です。
⑨正の数は絶対値が大きいほど大きく、負の数は絶対値が大きいほど小さくなります。

●「基準」を上回るかどうか、それが問題

31. Positive and negative numbers (question)

Question

① On the basis of Wednesday's production number of 150, the production number of goods made at a factory is shown in the table below.

基準 生産 数 品物 つくられた 示される 表 下の

② In this case, how many is the average daily production number?

この場合 いくつ 平均 1日の

Day of the week	Monday	Tuesday	Wednesday	Thursday	Friday	Saturday
Production number	+8	−12	0	−7	+14	+9

How to solve

③ For each day of the week, we think about the difference from Wednesday.

~について それぞれ 曜日 差

④ The total difference between from Monday to Saturday and Wednesday is (positive 8 minus 12 plus 0 minus 7 plus 14 plus 9 equals) 12.

合計の ~と…の間の (+8−12+0−7+14+9=)12

⑤ Because there are 6 days from Monday to

~なので ~がある

Saturday, 12 divided by 6 equals 2.
12÷6=2

⑥This expresses that the average production
表わす
number made each day is 2 greater than the basis of
それぞれの　～より大きい
Wednesday's 150.

⑦So, the average daily production number is 152.

Answer ⑧152 goods

31．正負の数（問題）

問　題　①ある工場での品物の生産数は、水曜日の生産数150個を基準にすると、下の表のようになりました。②このとき、1日の生産数の平均は何個ですか。

曜日	月	火	水	木	金	土
生産数	+8	−12	0	−7	+14	+9

解き方　③各曜日について、水曜日との差を考えます。④月曜日から土曜日までの水曜日との差の合計は、+8−12+0−7+14+9=12です。⑤月曜日から土曜日までは6日あるので、12÷6=2　⑥これは、1日の生産数の平均が基準の水曜日の150個より、2個多いことを表わしています。⑦よって、1日の生産数の平均は152個です。

答　え　⑧152個

32. How to express expressions with letters

① When we express products by using expressions
～するとき　表わす　積　式
with letters, we do it in the following ways.
文字　次の　方法

② As a times 5 equals $5a$, the multiplication symbol
～のように　$a\times5=5a$　乗法　記号
"×" is omitted.
省略される

③ In a product of a letter and a number, we write the
数
number before the letter.

④ The product of the same letters is expressed using
同じ　表わされる
an exponent as a times a equals a squared.
指数　$a\times a=a^2$

⑤ When we express quotients by using expressions
商
with letters, we do it in the following ways.

⑥ As a divided by 3 equals a over 3, the division
$a\div3=\frac{a}{3}$　除法
symbol "÷" is omitted.

⑦ The change from 1000 yen when buying 3 apples
おつり
at 1 apple x yen is expressed as 1000 minus $3x$
りんご1個 x 円で　$1000-3x$(円)
(yen).

⑧ The amount of juice each person gets when we divide y liters juice equally among 5 people is expressed as y over 5 liters.

amount 量　each それぞれの　divide 分ける　liters リットル(L)　equally 等しく　among 〜の間で

32. 文字式の表わし方

①文字を使った式で積を表わすときは、次のようにします。
② $a\times5=5a$ のように、乗法の記号「×」は省略します。
③文字と数の積は、数を文字の前に書きます。
④ $a\times a=a^2$ のように同じ文字の積は、指数を使って表わします。
⑤文字を使った式で商を表わすときは、次のようにします。
⑥ $a\div3=\frac{a}{3}$ のように、除法の記号「÷」は省略します。
⑦ 1個 x 円のりんごを3個買って、1000円出したときのおつりは、$1000-3x$(円)と表わせます。
⑧ yL のジュースを5人で等しく分けたときの1人分のジュースの量は、$\frac{y}{5}$(L)と表わせます。

● 「同類項」って、似たもの同士のこと？

33. How to calculate expressions with letters

① The expression $5x$ plus 2 is the sum of $5x$ and 2.
式 和

② In this case, the $5x$ and 2 connected by the addition symbol are called terms.
この場合 結ばれた 加法 記号 項

③ Also, in the term $5x$, the number 5 multiplied to the letter is called a coefficient of x.
また 数 かけられた 文字 呼ばれる 係数

④ Terms where the letter part are the same are called similar terms.
部分 同じ 同類項

⑤ Similar terms can be combined into a single term.
まとめられる 1つの

$$3x+2x=(3+2)x=5x$$

(3x plus 2x equals the quantity 3 plus 2, close quantity, times x equals 5x)

⑥ In multiplication of expressions with letters and numbers, we can multiply the coefficients and numbers by changing the order of the multiplication.
乗法 かける 順序

$$4a\times3=4\times3\times a=12a$$

(4a times 3 equals 4 times 3 times a equals 12a)

⑦ In the calculation of the following expression, we
計算 次の

use the distributive law and remove the parentheses
分配法則　取り除く　かっこ
and calculate similar terms together.
計算する　一緒に

$3(x+4)-2(2x-3)$ (3 times the quantity x plus 4, close quantity, minus 2 times the quantity $2x$ minus 3, close quantity)
$=3x+12-4x+6$ ($3x$ plus 12 minus $4x$ plus 6)
$=-x+18$ (negative x plus 18)

33. 文字式の計算

①$5x+2$の式は、$5x$と2の和です。
②このとき、加法の記号で結ばれた$5x$と2を項といいます。
③また、$5x$の項で、文字にかけられている数5をxの係数といいます。
④文字の部分が同じ項は同類項といいます。
⑤同類項は1つの項にまとめることができます。
⑥文字式と数の乗法は、かける順序を変えると、係数と数をかけることができます。
⑦次のような式の計算は、分配法則を使ってかっこをはずし、同類項をまとめて計算します。

●「マッチ棒」クイズなら、得意?!

34. Use of expressions with letters

Question

① How many matches make x squares when lining up matches?

いくつの　マッチ棒　x個の正方形　～のとき　並べる

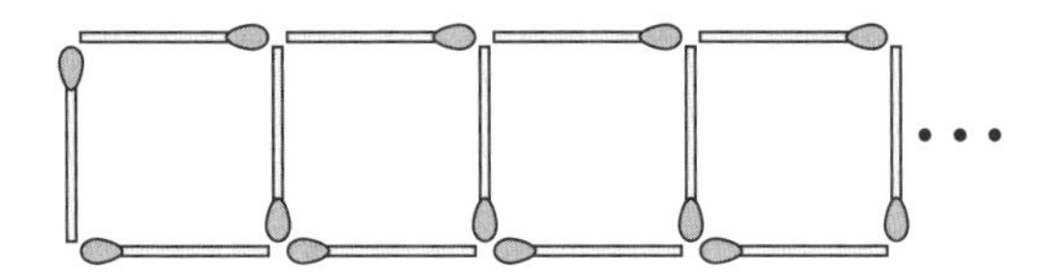

How to solve

② The number of matches when making a single square is 4.

数　1つの

③ Because the number of matches increases by 3 when there are 2 squares, the number of matches is (4 plus 3 equals) 7.

～なので　増える　～がある　(4+3=)7

④ Because the number of matches increases by 3 times the quantity x minus 1, close quantity, matches when there are x squares, the number of

3×(x−1)

matches is (4 plus 3 times the quantity x minus 1, close quantity, equals 4 plus $3x$ minus 3 equals) 1 plus $3x$.

⑤ Also if we consider that there are the furthest left single match and groups of 3 matches, (1 plus 3 times x equals) $3x$ plus 1.

Also：また　if：〜なら　consider：考える　furthest：一番遠い　groups：集まり

Answer ⑥ $3x$ plus 1

34. 文字式の利用

問　題　①マッチ棒を並べて、正方形を x 個つくるときのマッチ棒の本数は何本ですか。

(解き方)　②正方形が1個のときのマッチ棒の本数は4本です。

③正方形が2個のときはマッチ棒の本数が3本増えるので、マッチ棒の本数は $4+3=7$(本)です。

④正方形が x 個のとき、マッチ棒が $3\times(x-1)$(本)増えるので、マッチ棒の本数は、$4+3\times(x-1)=4+3x-3=1+3x$(本)です。

⑤また、一番左端の1本と3本のかたまりが x 個あると考えると、$1+3\times x=3x+1$(本)になります。

答　え　⑥ $3x+1$(本)

● 「因数分解」には四字熟語じゃないほうもある？

35. Prime Factorization

① When a natural number is expressed in the form as the product of several natural numbers, one by one, they are called factors of the original natural number.

～するとき 自然数 表わされる 形 ～として 積 いくつかの 1つずつ 呼ばれる 因数 もとの

② As for 32, it can be expressed as 4 times 8 or 2 times 16, so 2, 4, 8 and 16 are the factors of 32.

～についていえば 4×8 だから

③ Natural numbers which have no divisors other than 1 and itself are called prime numbers.

～がない 約数 ～以外 それ自体 素数

④ Prime numbers less than 20 are 2, 3, 5, 7, 11, 13, 17 and 19.

～より少ない

⑤ Decomposing natural numbers into the form of products of factors which are the prime numbers (prime factors) is called prime factorization.

分解すること 素因数 素因数分解

⑥ The prime factorization of 42 is 42 equals 2 times 3 times 7, and 60 is 60 equals 2 squared times 3 times 5.

$42=2\times3\times7$ $60=2^2\times3\times5$

⑦ When the numbers become large, if you divide by
大きく　わる
prime numbers you can do the prime factorization.

⑧ Dividing 420 in order from the smallest prime
わると　～から順に　いちばん小さい
number, it becomes 420 equals 2 squared times 3 times 5 times 7.

35．素因数分解

①自然数が、いくつかの自然数の積の形で表わされるとき、その１つ１つの数を、もとの自然数の因数といいます。
②32は、４×８、２×16と表わせるので、２、４、８、16は32の因数です。
③１とその数のほかに約数がない自然数を素数といいます。
④20以下の素数は、２、３、５、７、11、13、17、19です。
⑤自然数を素数である因数(素因数)の積の形に分解することを素因数分解といいます。
⑥42を素因数分解すると、$42 = 2 \times 3 \times 7$、60を素因数分解すると、$60 = 2^2 \times 3 \times 5$です。
⑦数が大きくなったときは、素数でわっていくことによって、素因数分解することができます。
⑧420を小さい素数から順にわっていくと、$420 = 2^2 \times 3 \times 5 \times 7$になります。

●2の2乗の「平方根」は？

36. Square roots

① When a certain number squared is x, that number is called the square root of x.

～するとき ある 数 2乗された 呼ばれる 平方根

② There are two square roots of 4, positive 2 and negative 2.

～がある +2 −2

③ The square root of 0 is only 0.

④ There is no square root of a negative number.

～がない 負の

⑤ Let's think about the square root of 2.

⑥ There is no whole number that becomes 2 when it is squared.

整数

⑦ The positive square root of 2 is one point four one four two one three five six ... and it is a decimal that continues without limit.

正の 1.41421356… 小数 続く ～なしに 限り

⑧ We use the radical sign $\sqrt{\ }$ and express this number as $\sqrt{2}$.

根号 表わす ～として

⑨ $\sqrt{2}$ is read "the square root of two".

読まれる

⑩ The square root of 2 is positive the square root of

$+\sqrt{2}$

2 and negative the square root of 2.
$-\sqrt{2}$

⑪ These two numbers are also integrated and
まとめられる
written $\pm\sqrt{2}$.
書かれる

36．平方根

①ある数を2乗すると x になるとき、ある数を x の平方根といいます。
②4の平方根は、＋2と－2の2つあります。
③0の平方根は0だけです。
④負の数には平方根はありません。
⑤2の平方根を考えてみましょう。
⑥2乗して2になる整数はありません。
⑦2の正の平方根は、1.41421356…と限りなく続く小数です。
⑧この数を、$\sqrt{\ }$（根号）を使って、$\sqrt{2}$と表わします。
⑨$\sqrt{2}$は「ルート2」と読みます。
⑩2の平方根は、$+\sqrt{2}$と$-\sqrt{2}$です。
⑪2つの数をまとめて$\pm\sqrt{2}$と書くこともあります。

●あなたの「ルート」はどこへ向かうルート？

37. Calculation of square roots

Question

① Do the calculations of the following.
計算 次のもの

② (1) $\sqrt{2} \times \sqrt{3}$ (The square root of 2 times the square root of 3)

③ (2) $2\sqrt{3} + 5\sqrt{3}$ (Two times the square root of 3 plus 5 times the square root of 3)

④ (3) $\sqrt{18} + \sqrt{8}$ (The square root of 18 plus the square root of 8)

How to solve

⑤ (1) The multiplication and division of expressions containing radical signs, are the products and quotients of the numerals inside the radical sign.
乗法 除法 式 を含む～ √‾ 積 商 数字 ～の中の

⑥ The square root of 2 times the square root of 3 equals the square root of 2 times 3 equals the square root of 6.
$\sqrt{2} \times \sqrt{3} = \sqrt{2 \times 3} = \sqrt{6}$

⑦ (2) Like $2\sqrt{3}$ and $5\sqrt{3}$, when the radical part is the same, they can be gathered together as 2 times the square root of 3 plus 5 times the square root of 3 equals 7 times the square root of 3.
～のように ～するとき √‾の 部分 同じ 集まる いっしょに ～として

⑧ (3) The square root of 18 plus the square root of 8

can be calculated (計算される) by simplifying (簡単にする) inside of each (それぞれの) radical sign. ⑨ The square root of 18 plus the square root of 8 equals 3 times the square root of 2 plus 2 times the square root of 2 equals 5 times the square root of 2.

Answer

⑩ (1) $\sqrt{6}$ (the square root of 6)　(2) $7\sqrt{3}$ (7 times the square root of 3)

(3) $5\sqrt{2}$ (5 times the square root of 2)

37. √‾の計算

問　題　①次の計算をしなさい。

②(1) $\sqrt{2}\times\sqrt{3}$　③(2) $2\sqrt{3}+5\sqrt{3}$

④(3) $\sqrt{18}+\sqrt{8}$

解き方　⑤(1)√‾を含む式の乗法と除法は、√‾の中の数字の積や商になります。

⑥$\sqrt{2}\times\sqrt{3}=\sqrt{2\times3}=\sqrt{6}$

⑦(2) $2\sqrt{3}$と$5\sqrt{3}$のように、√‾の部分が同じときは、$2\sqrt{3}+5\sqrt{3}=7\sqrt{3}$とまとめることができます。

⑧(3) $\sqrt{18}+\sqrt{8}$はそれぞれの√‾の中を簡単にすることで計算することができます。

⑨$\sqrt{18}+\sqrt{8}=3\sqrt{2}+2\sqrt{2}=5\sqrt{2}$

答　え　⑩(1) $\sqrt{6}$　(2) $7\sqrt{3}$　(3) $5\sqrt{2}$

● 「カッコ」があると、計算がうまくカッコつかない？

38. Multiplication and division of polynomials (expansion)

Question

① Expand the following expressions.
展開しなさい　次の～　式

② (1) $(x+4)(y+2)$ (the quantity x plus 4, close quantity, times the quantity y plus 2, close quantity)

③ (2) $(x+5)(x+4)$ (the quantity x plus 5, close quantity, times the quantity x plus 4, close quantity)

④ (3) $(x+6)^2$ (the square of the quantity x plus 6, close quantity)

⑤ (4) $(x+3)(x-3)$ (the quantity x plus 3, close quantity, times the quantity x minus 3, close quantity)

How to solve

⑥ (1) can be expanded like this,
展開される　～のように

$$(x+4)(y+2)=xy+2x+4y+8$$

(the quantity x plus 4, close quantity, times the quantity y plus 2, close quantity, equals xy plus $2x$ plus $4y$ plus 8)

⑦ (2) We expand by using the formula, the quantity x plus a, close quantity, times the quantity x plus b, close quantity equals x squared plus the quantity a plus b, close quantity, times x plus ab.
公式　$(x+a)(x+b)=$　$x^2+(a+b)x+ab$

⑧ (3) We expand by utilizing the formula, the square of the quantity a plus b, close quantity, equals
利用すること　$(a+b)^2=$

a squared plus $2ab$ plus b squared.
$a^2+2ab+b^2$

⑨(4) We expand by applying the formula, the quantity a plus b, close quantity, times the quantity a minus b, close quantity, equals a squared minus b squared.
applying: あてはめること　the quantity … close quantity, equals: $(a+b)(a-b)=$　a squared minus b squared: a^2-b^2

Answer

⑩(1) $xy+2x+4y+8$
(xy plus $2x$ plus $4y$ plus 8)

⑪(2) $x^2+9x+20$
(x squared plus $9x$ plus 20)

⑫(3) $x^2+12x+36$
(x squared plus $12x$ plus 36)

⑬(4) x^2-9
(x squared minus 9)

38. 多項式の乗除（展開）

問　題　①次の式を展開しなさい。

②(1) $(x+4)(y+2)$　　③(2) $(x+5)(x+4)$

④(3) $(x+6)^2$　　⑤(4) $(x+3)(x-3)$

解き方　⑥(1) $(x+4)(y+2)=xy+2x+4y+8$　と展開することができます。

⑦(2) 公式 $(x+a)(x+b)=x^2+(a+b)x+ab$ を使って展開します。

⑧(3) 公式 $(a+b)^2=a^2+2ab+b^2$ を利用して展開します。

⑨(4) 公式 $(a+b)(a-b)=a^2-b^2$ にあてはめて展開します。

答　え　⑩(1) $xy+2x+4y+8$　　⑪(2) $x^2+9x+20$

⑫(3) $x^2+12x+36$　　⑬(4) x^2-9

39. Factorization

Question

① Let's factorize the following expressions.
因数分解する　次の　式

② (1) $mx+my$
(mx plus my)

③ (2) $x^2+10x+25$
(x squared plus $10x$ plus 25)

④ (3) $x^2+7x+12$
(x squared plus $7x$ plus 12)

⑤ (4) x^2-16
(x squared minus 16)

How to solve

⑥ Expressing a polynomial in the form of a product of some factors is called "factorize".
表わすこと　多項式　形　積　因数　呼ばれる

⑦ In (1), because there is a common factor m in each term, pull it out of the parenthesis.
～なので　共通の　それぞれの　項　引く　～から　かっこ

⑧ In (2), utilize the formula a squared plus $2ab$ plus b squared equals the quantity a plus b, close quantity, squared.
利用する　公式　$a^2+2ab+b^2=(a+b)^2$

⑨ In (3), use the formula x squared plus the quantity a plus b, close quantity, times x plus ab equals the quantity x plus a, close quantity, times the quantity x plus b, close quantity.
$x^2+(a+b)x+ab=(x+a)(x+b)$

⑩In (4), apply the formula a squared minus b squared equals the quantity a plus b, close quantity, times the quantity a minus b, close quantity.

apply：適用する
a squared minus b squared equals the quantity a plus b, close quantity, times the quantity a minus b, close quantity：$a^2-b^2=(a+b)(a-b)$

Answer

⑪ (1) $m(x+y)$ (m times the quantity x plus y, close quantity)

⑫ (2) $(x+5)^2$ (the square of the quantity x plus 5, close quantity)

⑬ (3) $(x+3)(x+4)$ (the quantity x plus 3, close quantity times the quantity x plus 4, close quantity)

⑭ (4) $(x+4)(x-4)$ (the quantity x plus 4, close quantity times the quantity x minus 4, close quantity)

39. 因数分解

問　題　①次の式を因数分解しましょう。

②(1) $mx+my$　　③(2) $x^2+10x+25$

④(3) $x^2+7x+12$　　⑤(4) x^2-16

解き方　⑥多項式をいくつかの因数の積の形に表わすことを、因数分解するといいます。

⑦(1)は、各項に共通な因数 m があるので、かっこの外にくくり出します。

⑧(2)は、$a^2+2ab+b^2=(a+b)^2$ の公式を利用します。

⑨(3)は、$x^2+(a+b)x+ab=(x+a)(x+b)$ の公式を使います。

⑩(4)は、$a^2-b^2=(a+b)(a-b)$ の公式にあてはめます。

答　え　⑪(1) $m(x+y)$　　⑫(2) $(x+5)^2$

⑬(3) $(x+3)(x+4)$　　⑭(4) $(x+4)(x-4)$

●「わからない」もの同士が＝(イコール)だなんて……

40. How to solve linear equations

Question

① Let's solve the following linear equation.
解く　次の～　1次方程式

② $4x-8=-5x-17$ (4x minus 8 equals negative 5x minus 17)

How to solve

③ An expression which expresses the relationship of quantity by using an equal sign is called an equality.
式　表わす　関係　数量　等号　呼ばれる　等式

④ An equality including letters is called an equation.
含む　文字　方程式

⑤ The value of the letters making up the equation is called the solution of the equation.
値　成り立たせる　解

⑥ Finding the solution of an equation is called "solve the equation."
求めること

⑦ Linear equations are calculated by transposing the terms with letters to the left side and terms with only numbers to the right side.
計算される　移項すること　項　辺　数

$4x-8=-5x-17$ (4*x* minus 8 equals negative 5*x* minus 17)

$4x+5x=-17+8$ (4*x* plus 5*x* equals negative 17 plus 8)

$9x=-9$ (9*x* equals negative 9)

$x=-1$ (*x* equals negative 1)

Answer ⑧ $x=-1$ (*x* equals negative 1)

40．1次方程式の解き方

問　題　①次の1次方程式を解きましょう。
②$4x-8=-5x-17$

(解き方) ③等号を使って、数量の関係を表わした式を等式といいます。④文字を含む等式を、方程式といいます。⑤方程式を成り立たせる文字の値を、その方程式の解(かい)といいます。⑥方程式の解を求めることを、「方程式を解く」といいます。⑦1次方程式は、文字のある項を左辺に、数字だけの項を右辺に移項して計算します。

$$4x-8=-5x-17$$
$$4x+5x=-17+8$$
$$9x=-9$$
$$x=-1$$

答　え ⑧ $x=-1$

●足らないものや「余る」もののことは忘れたい?!

41. Linear equations word problem (too many or too few)

Question

① Some pieces of candy are given to some children.
数個の　飴　与えられる
② When each child gets 5 pieces of candy, there are
～するとき　それぞれの　～がある
3 pieces too many. ③ When each child gets 6 pieces
多い
of candy, there are 5 pieces too few. ④ Find an answer
少ない　求めなさい
to the number of children and the number of candy
数
pieces in this case.
この場合

How to solve

⑤ Assuming that the number of children is x, the
仮定すると
number of candy pieces can be expressed in two ways.
表わされる　方法
⑥ One is, from "When each child gets 5 pieces of

candy, there are 3 pieces too many," (5x plus 3) pieces.
$5x+3$
⑦ Another is, from "When each child gets 6 pieces of
もう1つ
candy, there are 5 pieces too few," (6x minus 5) pieces.
$6x-5$
⑧ Because these two equations express the same
～なので　方程式　表わす　同じ
number of candy pieces, it becomes 5x plus 3 equals
$5x+3=6x-5$

$6x$ minus 5.

⑨ Solving this equation,
解くと

$5x-6x=-5-3$ (5x minus 6x equals negative 5 minus 3)

$-x=-8$ (negative x equals negative 8)

$x=8$ (x equals 8)

⑩ Therefore the number of children is 8. ⑪ The number
よって
of candy pieces is (5 times 8 plus 3 equals) 43.
5×8+3=43

Answer ⑫ 8 children, 43 pieces of candy

41．1次方程式の文章題（過不足）

問　題 ①何人かの子どもに飴（あめ）を配ります。②1人に飴を5個ずつ配ると、3個あまります。③1人に飴を6個ずつ配ると、5個足りません。④このとき、子どもの人数と飴の個数を求めなさい。

解き方 ⑤子どもの人数を x 人と仮定すると、飴の個数は、2通りの式で表わすことができます。⑥1つは、「1人に飴を5個ずつ配ると、3個あまる」ことから、$(5x+3)$個。⑦もう1つは、「1人に飴を6個ずつ配ると、5個足りない」ことから、$(6x-5)$個。⑧この2つの式は、同じ個数を表わしているので、$5x+3=6x-5$となります。⑨これを解くと、

$5x-6x=-5-3$　　$-x=-8$　　$x=8$

⑩よって、子どもの人数は8人。⑪飴の個数は、$5\times8+3=43$(個)。

答　え ⑫子どもの人数8人、飴の個数43個

●2つ「未知数」があっても、2つ式があれば……

42. How to solve simultaneous equations (addition and subtraction method)

Question

①Let's solve the following simultaneous equations by the addition and subtraction method.

解く　次の〜　連立方程式　加減法

$$\begin{cases} 4x-3y=5 & \text{(4x minus 3y equals 5)} \cdots\cdots(1) \\ 3x+2y=8 & \text{(3x plus 2y equals 8)} \cdots\cdots(2) \end{cases}$$

How to solve

② A combination of 2 or more equations is called a simultaneous equation.

組み合わせ　2つかそれ以上　方程式　呼ばれる

③ The way to erase the letters by making equal the absolute values of the coefficients of the either letters of simultaneous equation and adding or subtracting is called the addition and subtraction method. ④ For this problem, to make equal the absolute value of the coefficient of y, we multiply the equation (1) by 2 and the equation (2) by 3.

方法　消去する　文字　〜にする　等しい　絶対値　係数　どちらかの　加えること　引くこと　〜について　かける

$$\begin{array}{rrl} & 8x-6y=10 & \text{(8x minus 6y equals 10)} \cdots\cdots(1)\times 2 \\ +) & 9x+6y=24 & \text{(plus 9x plus 6y equals 24)} \cdots\cdots(2)\times 3 \\ \hline & 17x=34 & \text{(17x equals 34)} \end{array}$$

$x=2$ (x equals 2) ……(3)

⑤ Substituting (3) into (1),
代入して　〜に

$4\times 2-3y=5$ (4 times 2 minus 3y equals 5)

$-3y=-3$ (negative 3y equals negative 3)

$y=1$ (y equals 1)

Answer ⑥ $x=2,\ y=1$ (x equals 2, y equals 1)

42. 連立方程式の解き方（加減法）

問　題　①次の連立方程式を、加減法で解きましょう。

$$\begin{cases} 4x-3y=5 & \cdots\cdots(1) \\ 3x+2y=8 & \cdots\cdots(2) \end{cases}$$

解き方　②2つ以上の方程式を組み合わせたものを、連立方程式といいます。③連立方程式のどちらかの文字の係数の絶対値を等しくして、加えたり引いたりして、文字を消去する解き方を加減法といいます。④この問題の場合、yの係数の絶対値を等しくするために、(1)の式を2倍、(2)の式を3倍します。

$$\begin{array}{rl} 8x-6y=10 & \cdots\cdots(1)\times 2 \\ +)\ \ 9x+6y=24 & \cdots\cdots(2)\times 3 \\ \hline 17x\qquad\ =34 & \\ x=2 & \cdots\cdots(3) \end{array}$$

⑤(3)を(1)に代入すると、

$4\times 2-3y=5$

$-3y=-3$

$y=1$

答　え ⑥ $x=2$、$y=1$

●当てずっぽうに、とりあえず、「あてはめる」のはダメ?!

43. How to solve simultaneous equations (substitution method)

Question

① Let's solve the following simultaneous equations by the substitution method.

解く　次の〜　連立方程式　代入法

$$\begin{cases} 3x+2y=-1 & \text{(3x plus 2y equals negative 1)} \cdots\cdots(1) \\ x-4y=9 & \text{(x minus 4y equals 9)} \cdots\cdots(2) \end{cases}$$

How to solve

② As with the addition and subtraction method, the substitution method also solves simultaneous equations by erasing a single letter.

〜と同じように　加減法　代入法　〜もまた　消去すること　1つの　文字

③ In the substitution method, by substituting one equation into the other, a letter is erased.

代入すること　〜に　もう一方　消去される

④ For this problem, if we rearrange equation (2), it becomes x equals $4y$ plus 9 ……(3).

〜について　〜なら　整理する　$x=4y+9$

⑤ Substituting equation (3) into equation (1),

$$3(4y+9)+2y=-1 \quad \text{(3 times the quantity 4y plus 9, close quantity, plus 2y equals negative 1)}$$

$$12y+27+2y=-1 \quad \text{(12y plus 27 plus 2y equals negative 1)}$$

$14y=-28$ (14y equals negative 28)

$y=-2$ (y equals negative 2) ……(4)

⑥ Substituting equation (4) into equation (3),

$x=4\times(-2)+9$ (x equals 4 times negative 2 plus 9)

$x=1$ (x equals 1)

Answer ⑦ $x=1,\ y=-2$ (x equals 1, y equals negative 2)

43. 連立方程式の解き方（代入法）

問　題 ①次の連立方程式を、代入法で解きましょう。

$$\begin{cases} 3x+2y=-1 & \cdots\cdots(1) \\ x-4y=9 & \cdots\cdots(2) \end{cases}$$

解き方 ②代入法も加減法と同じように、1つの文字を消去して連立方程式を解きます。③代入法では、一方の式を他方の式に代入することで、文字を消去します。④この問題の場合、(2)の式を整理すると、$x=4y+9$ ……(3)となります。⑤(3)の式を(1)の式に代入すると、

$3(4y+9)+2y=-1$

$12y+27+2y=-1$

$14y=-28$

$y=-2$ ……(4)

⑥(4)を(3)に代入すると、

$x=4\times(-2)+9$、$x=1$

答　え ⑦ $x=1$、$y=-2$

● 「塩から」くて、飲めませんが……

44. Simultaneous equations word problem (salt solutions)

Question

① A 10% concentration salt solution and a 15%
濃度 食塩水
concentration salt solution were mixed to make 600
混ぜられた
grams of 12% salt solution. ② How many grams of
グラム(g) 何グラム
each of the 2 kinds of salt solution were mixed
それぞれ 種類
together?
いっしょに

How to solve

③ We assume, x grams of the 10% salt solution and y
仮定する
grams of the 15% salt solution are mixed. ④ Because
〜なので
the weight of the mixed salt solution is 600 grams,
重さ 混ぜられた
the equation becomes x plus y equals 600.
方程式 $x+y=600$
⑤ In addition, because the weight of the salt can
さらに
be found by "the weight of the salt solution times
求められる かける
the concentration of salt solution", the equation

becomes 10 over 100 times x plus 15 over 100 times
$\frac{10}{100}x+\frac{15}{100}y=600\times\frac{12}{100}$
y equals 600 times 12 over 100. ⑥ We solve these 2
解く

equations as a simultaneous equation.
~として 連立方程式

$$\begin{cases} x+y=600\cdots(1) \\ \dfrac{10}{100}x+\dfrac{15}{100}y=600\times\dfrac{12}{100}\cdots(2) \end{cases}$$

(x plus y equals 600)

(10 over 100 times x plus 15 over 100 times y equals 600 times 12 over 100)

$x=360$ (x equals 360), $y=240$ (y equals 240)

Answer ⑦10% salt solution 360 grams, 15% salt solution 240 grams

44. 連立方程式の文章題（食塩水）

問　題　①濃度が10%の食塩水と15%の食塩水を混ぜて、12%の食塩水を600gつくりました。②2種類の食塩水をそれぞれ何gずつ混ぜましたか。

解き方　③10%の食塩水をxg、15%の食塩水をyg混ぜたと仮定します。④混ぜてできた食塩水の重さは600gなので、$x+y=600$となります。⑤また、食塩の重さは、「食塩水の重さ×食塩水の濃度」で求めることができるので、$\dfrac{10}{100}x+\dfrac{15}{100}y=600\times\dfrac{12}{100}$となります。
⑥これらの2つの式を連立方程式として解きます。

$$\begin{cases} x+y=600\cdots(1) \\ \dfrac{10}{100}x+\dfrac{15}{100}y=600\times\dfrac{12}{100}\cdots(2) \end{cases}$$

$x=360$、$y=240$

答　え　⑦10%の食塩水360g、15%の食塩水240g

●「2次方程式」は＝(イコール)の右と左を……？

45. How to solve quadratic equations

Question

① Let's solve the following equations.
解く 次の〜 方程式

② (1) $x^2-4=14$ (x squared minus 4 equals 14)

③ (2) $x^2-x-56=0$ (x squared minus x minus 56 equals 0)

How to solve

④ When rearranged by transposing, the equation that can be transformed into the form, (quadratic expression) equals zero is called a quadratic equation.
〜するとき 整理される 移項すること 変形される 〜に 形 (2次式)＝0 呼ばれる

⑤ (1) When transposed, x squared equals 18.
移項される

⑥ x squared equals 18 shows that x is the square root of 18.
示す 平方根

$x^2=18$ (x squared equals 18)

$x=\pm\sqrt{18}$ (x equals positive or negative the square root of 18)

$x=\pm3\sqrt{2}$ (x equals positive or negative 3 times the square root of 2)

⑦ (2) We solve the left side by factorizing, using the
辺 因数分解すること

formula, x squared plus the quantity a plus b, close
公式　$x^2+(a+b)x+ab=(x+a)(x+b)$
quantity, times x plus $a\,b$ equals the quantity x plus

a, close quantity, times the quantity x plus b, close

quantity.

$x^2-x-56=0$ (x squared minus x minus 56 equals 0)

$(x-8)(x+7)=0$ (the quantity x minus 8, close quantity, times the quantity x plus 7, close quantity, equals 0)

$x=8, -7$ (x equals 8, negative 7)

Answer

⑧(1) $x=\pm 3\sqrt{2}$ (x equals positive or negative 3 times the square root of 2)

(2) $x=8, -7$ (x equals 8, negative 7)

45. 2次方程式の解き方

問　題　①次の方程式を解きましょう。

②(1) $x^2-4=14$　　③(2) $x^2-x-56=0$

解き方　④移項して整理すると、（2次式）＝0の形に変形できる方程式を、2次方程式といいます。

⑤(1)移項すると、$x^2=18$となります。

⑥ $x^2=18$は、xが18の平方根であることを示しています。

⑦(2) $x^2+(a+b)x+ab=(x+a)(x+b)$の公式を使って、左辺を因数分解して解きます。

答　え　⑧(1) $x=\pm 3\sqrt{2}$　　(2) $x=8$、-7

46. Quadratic equations and the solution formula

Question

①Let's solve the following quadratic equation by using the solution formula.
解く　次の〜　2次方程式
解の公式

②$2x^2-5x+1=0$ (2x squared minus 5x plus 1 equals 0)

How to solve

③The solution of the quadratic equation, $a$$x$ squared plus bx plus c equals 0 is, x equals negative b plus or minus the square root of b squared minus 4ac over 2a.
$ax^2+bx+c=0$
$x=\frac{-b\pm\sqrt{b^2-4ac}}{2a}$

④Solutions can be found by substituting the values of a, b and c into this equation.
求められる　代入すること
値　〜に

⑤The equation, x equals negative b plus or minus the square root of b squared minus 4ac over 2a, is called the solution formula.
呼ばれる

⑥From 2x squared minus 5x plus 1 equals 0, we substitute a equals 2, b equals negative 5 and c
代入する

equals 1 into the solution formula.

$$x=\frac{-(-5)\pm\sqrt{(-5)^2-4\times2\times1}}{2\times2}$$ (x equals negative negative 5 plus or minus the square root of negative 5 squared minus 4 times 2 times 1 over 2 times 2)

$$=\frac{5\pm\sqrt{17}}{4}$$ (equals 5 plus or minus the square root of 17 over 4)

Answer ⑦ $x=\frac{5\pm\sqrt{17}}{4}$ (x equals 5 plus or minus the square root of 17 over 4)

46. ２次方程式と解の公式

問　題　①次の２次方程式を、解の公式を使って解きましょう。

②$2x^2-5x+1=0$

解き方　③２次方程式 $ax^2+bx+c=0$ の解は、

$x=\frac{-b\pm\sqrt{b^2-4ac}}{2a}$ です。

④この式に a、b、c の値を代入することで解を求めることができます。

⑤ $x=\frac{-b\pm\sqrt{b^2-4ac}}{2a}$ の式を解の公式といいます。

⑥$2x^2-5x+1=0$より、$a=2$、$b=-5$、$c=1$を解の公式に代入します。

$$x=\frac{-(-5)\pm\sqrt{(-5)^2-4\times2\times1}}{2\times2}$$

$$=\frac{5\pm\sqrt{17}}{4}$$

答　え ⑦ $x=\frac{5\pm\sqrt{17}}{4}$

How to express large numbers in English

英語では大きい数をどのように表わすのか

① In Japanese, 10 cubed is called "*sen*," 10 to the fourth power is "*man*," 10 to the eighth power is "*oku*," and 10 to the twelfth power is "*cho*." ② How are large numbers expressed in English? ③ In English, 10 cubed which is "*sen*" in Japanese, is called thousand. ④ However, English does not have an equal word to the Japanese word "*man*." ⑤ In English, "*ichi man*" is expressed as "ten thousand." ⑥ In English, the next larger separation after thousand is million which is expressed as 10 to the sixth power. ⑦ This is equal to "*hyaku man*" in Japanese. ⑧ The next larger separation is billion. ⑨ This is ten to the ninth power which in other words equals "*jyu oku*" in Japanese. ⑩ An even larger separation is trillion. ⑪ This is expressed as ten to the twelfth power. ⑫ It equals "*iccho*" in Japanese. ⑬ In English, "1,000" is expressed as thousand, "1,000,000" as million, "1,000,000,000"as billion, and "1,000,000,000,000" as trillion.

---

①日本語では、10^3を「千」、10^4を「万」、10^8を「億」、10^{12}を「兆」といいます。②英語では大きい数をどのように表わすのでしょうか。③英語では、日本語の「千」にあたる10^3を thousand といいます。④しかし、日本語の「万」にあたる語は、英語にはありません。⑤英語では、「一万」を「ten thousand」と表わすのです。⑥英語で thousand の次に大きい区切りは、10^6を表わす million です。⑦これは、日本語では「百万」にあたります。⑧その次に大きい区切りは billion です。⑨これは10^9、つまり日本語の「十億」にあたります。⑩さらに大きい区切りは trillion です。⑪これは10^{12}を表わします。⑫日本語では「一兆」にあたります。⑬英語では、「1,000」は thousand、「1,000,000」は million、「1,000,000,000」は billion、「1,000,000,000,000」は trillion と表わすのです。

Chapter 3 Functions

第3章　関　数

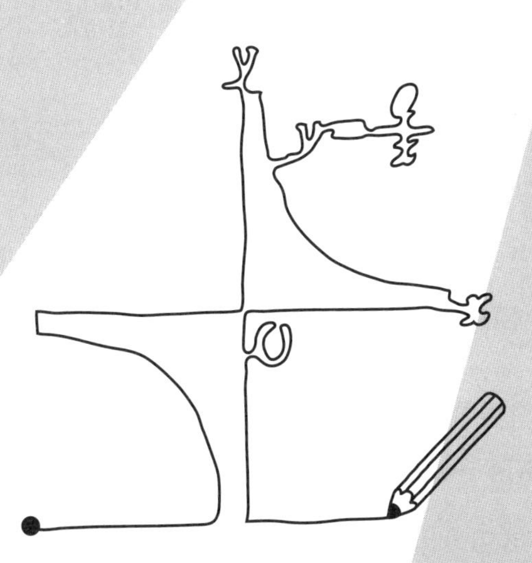

47. Proportion and inverse proportion

① When the value of x is doubled, tripled··· and
～するとき　値　2倍される　3倍される
along with it the value of y is also doubled, tripled···,
～といっしょに
we say y is in proportion to x.
～に比例して

② In this case, the relationship of x and y is expressed,
この場合　関係　表わされる
using the proportionality constant a, in the equation
比例定数　方程式
y equals ax.
$y=ax$

③ The graph of the relationship
グラフ
of proportion y equals ax becomes,
like in Figure 1, a straight line
～のように　図　直線
passing through the origin O.
通る　原点

Figure 1
(図1)

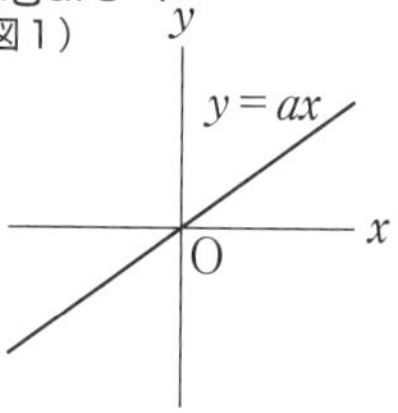

④ When the value of x is doubled, tripled··· and
along with it the value of y is halved, multiplied by
$\frac{1}{2}$にされる　$\frac{1}{3}$でかけられる
1 over 3···, we say y is in inverse proportion to x.
～に反比例して

⑤ In this case, the relationship of x and y is expressed, using the proportionality constant a, in the equation y equals a over x.
$y=\frac{a}{x}$

⑥ The graph of the relationship of inverse proportion y equals a over x becomes, like in Figure 2, a curve line (曲線) called (呼ばれる) a hyperbolic curve (双曲線). ⑦ There are (〜がある) the variables (変数) x and y which vary (変わる) along with each other (お互い), when the value of x is decided (決められる) and there is only one value of y, y is called a function (関数) of x.

Figure 2
(図2)

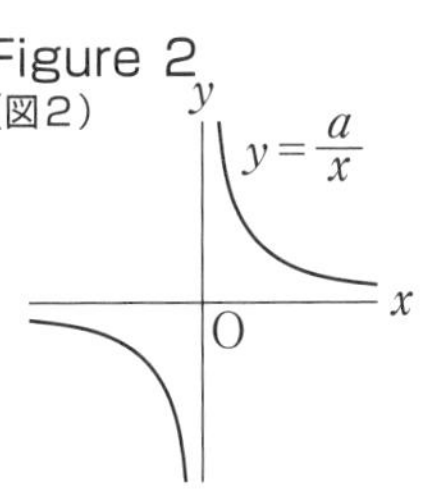

47. 比例と反比例

① x の値が2倍、3倍、…になると、それにともなって y の値も2倍、3倍、…になるとき、y は x に比例するといいます。②このとき、x、y の関係は比例定数 a を使った $y=ax$ の式で表わされます。③比例の関係 $y=ax$ のグラフは、図1のように、原点Oを通る直線になります。④ x の値が2倍、3倍、…になると、それにともなって y の値が $\frac{1}{2}$ 倍、$\frac{1}{3}$ 倍、…になるとき、y は x に反比例するといいます。⑤このとき、x、y の関係は比例定数 a を使った $y=\frac{a}{x}$ の式で表わされます。⑥反比例の関係 $y=\frac{a}{x}$ のグラフは、図2のように、双曲線と呼ばれる曲線になります。⑦ともなって変わる変数 x、y があり、x の値を決めると y の値がただ1つに決まるとき、y は x の関数であるといいます。

●倍返しって……、「比例」だな？

48. Use of proportion (word problem)

Question

①The weight of 15 sheets of some copier paper is
重さ　15枚の　ある〜　コピー用紙
51 grams. ②Express y as the expression of x,
グラム(g)　表わしなさい　〜として　式
thinking y grams as the weight of x sheets of copier
考えて
paper. ③Also, how many grams is the weight of 215
また　何グラム
sheets of the copier paper?

How to solve

④Because y weight of copier paper is in proportion
〜なので　〜に比例して
to x numbers of sheets, it can be expressed with the
数　表わされる
equation y equals ax. ⑤Because the weight of 15
方程式　$y=ax$
sheets of copier paper is 51 grams, substitute
代入する
x equals 15 and y equals 51 into y equals ax.
$x=15$　$y=51$

$51=a\times 15$ (51 equals a times 15)

$15a=51$ (15a equals 51)

$a=\frac{17}{5}$ (a equals 17 over 5)

⑥Therefore, the equation can be expressed as
よって

y equals 17 over 5 times x. ⑦If we substitute x equals
$y=\frac{17}{5}x$　～なら

215 into this equation, y equals 17 over 5 times 215
$y=\frac{17}{5}\times215$

becomes y equals 731. ⑧The weight of 215 sheets of copier paper is 731 grams.

Answer ⑨equation : y equals 17 over 5 times x, weight : 731 grams

48. 比例の利用（文章題）

問　題　①あるコピー用紙15枚分の重さをはかると、51gでした。②コピー用紙 x 枚の重さを yg として、y を x の式で表わしなさい。③また、このコピー用紙215枚分の重さは何 g ですか。

解き方　④コピー用紙の重さ y は枚数 x に比例するので、$y=ax$ の式で表わされます。⑤コピー用紙15枚分の重さが51gだから、$y=ax$ に $x=15$、$y=51$を代入します。

$51=a\times15$

$15a=51$

$a=\frac{17}{5}$

⑥よって、$y=\frac{17}{5}x$ の式で表わされます。⑦この式に、$x=215$を代入すると、$y=\frac{17}{5}\times215$、$y=731$となります。⑧コピー用紙215枚分の重さは731gです。

答　え　⑨式：$y=\frac{17}{5}x$、重さ：731g

●「右肩上がり」って、いいよね？

49. Use of proportion (graph)

Question

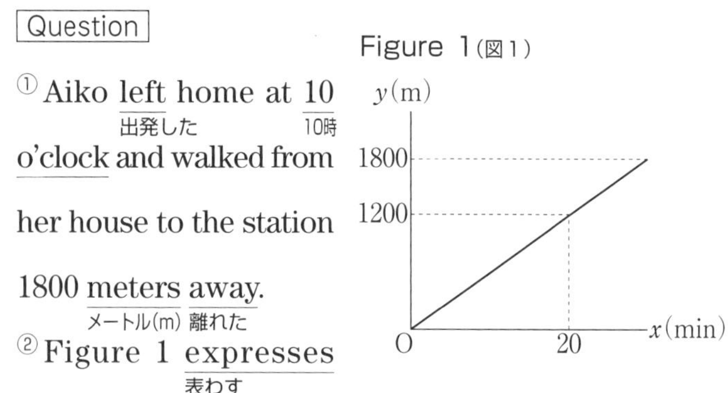

① Aiko left (出発した) home at 10 o'clock (10時) and walked from her house to the station 1800 meters (メートル(m)) away (離れた).

② Figure 1 expresses (表わす) the relationship (関係) of the time from when (～したとき) Aiko left home and the distance (道のり) from her house.

③ (1) Express y as (～として) the expression (式) of x, thinking (～として考えて) x minutes as the time after Aiko left home and thinking y meters as the distance from her house.

④ (2) Find (求めなさい) how many minutes (何分) after she left home did Aiko arrive (着く) at the station.

How to solve

⑤ Figure 1 is a straight line (直線) passing through (～を通る) the origin (原点), therefore (よって) y is in proportion to (～に比例する) x.

⑥We substitute (代入する) x equals 20 (x=20) and y equals 1200 into (～に) y equals ax.

⑦From (～より) 1200 equals $20a$, a equals 60.

⑧Because (～なので) the station is 1800 meters away from her house, we substitute y equals 1800 into y equals $60x$.

⑨From 1800 equals $60x$, x equals 30, it becomes 30 minutes after.

Answer ⑩(1) y equals $60x$ ⑪(2)30 minutes after

49. 比例の利用（グラフ）

問 題 ①あいこさんは10時に家を出発して、家から1800m 離れた駅まで歩いて行きました。②図1は、あいこさんが家を出発してからの時間と家からの道のりの関係を表わしたものです。③(1)あいこさんが家を出発してからの時間 x 分と、家からの道のり ym について、y を x の式で表わしなさい。④(2)あいこさんが駅に着いたのは、家を出発してから何分後かを求めなさい。

解き方 ⑤図1は原点を通る直線だから、y は x に比例します。⑥$y=ax$ に、$x=20$、$y=1200$を代入します。⑦$1200=20a$ より、$a=60$となります。⑧駅は家から1800m 離れているので、$y=60x$ に、$y=1800$を代入します。⑨$1800=60x$、$x=30$より、30分後となります。

答 え ⑩(1) $y=60x$ ⑪(2) 30分後

●「座標」って、裏があるんだよね……

50. Proportion and inverse proportion (graph)

Question

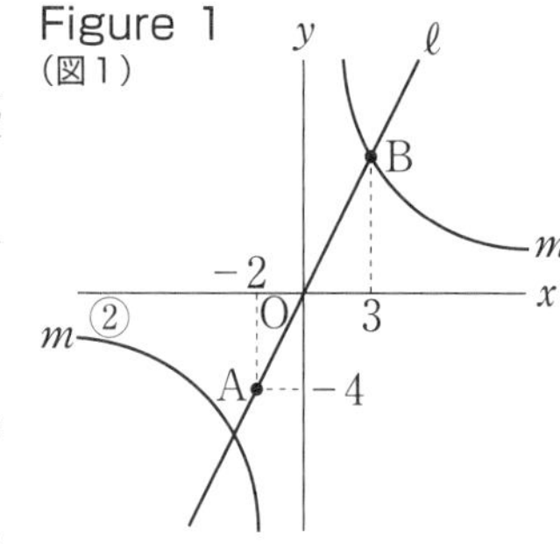

① In Figure 1, the straight line ℓ is the graph of y equals ax.
図　直線　$y=ax$

② Curve line m is the graph of y equals b over x.
曲線　$y=\frac{b}{x}$

③ Point A is a point on straight line ℓ.
点

④ Point B is the intersection point of curve line m and straight line ℓ.
交点

⑤ Find the value of a and b when the coordinates of point A are (negative 2, negative 4) and the x coordinate of point B is 3.
求めなさい　値　〜するとき　座標　$(-2, -4)$

How to solve

⑥ In order to find a, we substitute x equals negative 2 and y equals negative 4 into y equals ax.
〜するために　代入する　$x=-2$　〜に

⑦ From negative 4 equals negative $2a$, a equals 2.
〜より

⑧ Because point B is a point on straight line ℓ, we substitute x equals 3 into y equals $2x$.
～なので

⑨ y equals 2 times 3 equals 6.
$y=2\times3=6$

⑩ The coordinates of point B are (3, 6).

⑪ Because point B is a point on curve line m, in order to find b, we substitute x equals 3 and y equals 6 into y equals b over x.

⑫ From 6 equals b over 3, b equals 18.

Answer ⑬ a equals 2, b equals 18

50. 比例と反比例（グラフ）

問　題 ①図１で、直線 ℓ は $y=ax$ のグラフです。②曲線 m は $y=\frac{b}{x}$ のグラフです。③点Ａは直線 ℓ 上の点です。④点Ｂは直線 ℓ と曲線 m の交点です。⑤点Ａの座標が$(-2, -4)$、点Ｂの x 座標が３のとき、a、b の値を求めなさい。

解き方 ⑥ a を求めるために、$y=ax$ に、$x=-2$、$y=-4$を代入します。⑦$-4=-2a$ より、$a=2$となります。⑧点Ｂは直線 ℓ 上の点だから、$y=2x$ に、$x=3$を代入します。⑨$y=2\times3=6$となります。⑩点Ｂの座標は、$(3, 6)$です。⑪点Ｂは曲線 m 上の点だから、b を求めるために、$y=\frac{b}{x}$ に $x=3$、$y=6$を代入します。⑫$6=\frac{b}{3}$ より、$b=18$です。

答　え ⑬ $a=2$、$b=18$

●「単純」なことって、スッキリ表わせるんだ

51. Linear functions (introduction)

①With the two variables of x, y, when the value of x is
〜があって　変数　…するとき　値
decided and the value of y is then decided accordingly,
決められる　そのとき　それに応じて
it is said that y is a function of x.②When y is
…といわれる　関数
expressed by the linear expression of x, y is called the
表わされる　リニア 1次元の　式　〜と呼ばれる
linear function of x.③In general, the linear function
1次関数　一般に
is expressed by the equation y equals $a\,x$ plus b,
$y=ax+b$
using constants, a and b.④In the linear function y
定数
equals $a\,x$ plus b, the ratio of change (the increased
変化の割合　増加量
amount of x to the increased amount of y) is
constant and equal to a.⑤The graph of the linear
一定な　等しい　グラフ
function y equals $a\,x$ plus b becomes a straight line
直線
with a slope of a and an intercept of b.⑥A slope
傾き　切片
expresses how much y changed when the value of
どれだけ
x increased by 1.⑦An intercept expresses the value
増加した
of y when x equals 0.⑧For example, Figure 1 is the
図
graph of y equals $2x$ plus 1.

⑨ If the ratio of change is 2, the slope of the graph also becomes 2, and when it goes to the right 1, it goes up 2.

~なら

Figure 1
（図1）

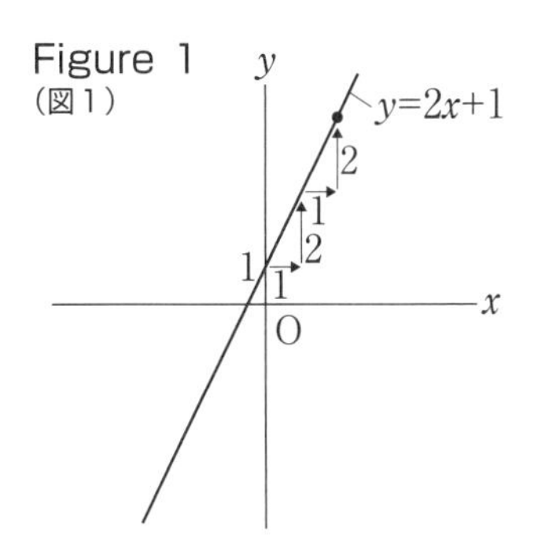

51. 1次関数（導入）

①2つの変数 x、y があって、x の値が決まると、それに対応して y の値が1つに決まるとき、y は x の関数であるといいます。②y が x の1次式で表わされるとき、y は x の1次関数であるといいます。③1次関数は、一般に、定数 a、b を使って $y=ax+b$ の式で表わされます。④1次関数 $y=ax+b$ では変化の割合（x の増加量に対する y の増加量）は一定で、a に等しいです。⑤1次関数 $y=ax+b$ のグラフは、傾きが a で、切片が b の直線になります。⑥傾きは、x の値が1増えたとき、y がどれだけ変化したかを表わします。⑦切片は、$x=0$のときの y の値を表わします。
⑧たとえば、図1は $y=2x+1$のグラフです。⑨変化の割合が2の場合、グラフの傾きも2となり、右へ1進むと、上へ2進みます。

52. Problem of finding an equation of linear functions

Question

① y is the linear function of x, and its graph passes through 2 points (negative 1, 5) and (2, negative 4).

1次関数　～を通る　点　(−1, 5)

② Find the equation of this linear function.

求めなさい　方程式

How to solve

③ We think the equation of the linear function to find as y equals ax plus b.

～として考える　$y=ax+b$

④ Because the graph passes through the 2 points (negative

～なので

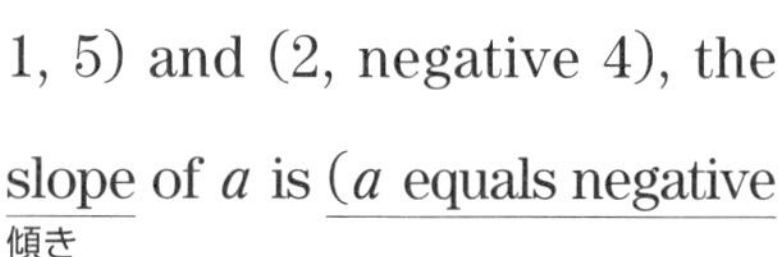

1, 5) and (2, negative 4), the slope of a is (a equals negative 4 minus 5 over 2 minus negative

傾き

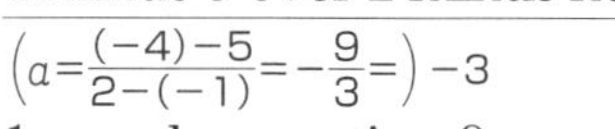

$\left(a=\frac{(-4)-5}{2-(-1)}=-\frac{9}{3}=\right)-3$

1 equals negative 9 over 3 equals) negative 3. ⑤ So, the equation is expressed as y equals negative $3x$ plus b. ⑥ Because the graph passes through point

よって　表わされる　～として

(negative 1, 5), we substitute x equals negative 1 and
代入する
y equals 5 into y equals negative $3x$ plus b. ⑦ From
～に ～より
5 equals negative 3 times negative 1 plus b, b equals 2.
$5=-3\times(-1)+b$
⑧ So, the equation we are finding becomes y equals
求めている
negative $3x$ plus 2. ⑨ Because this graph also passes
～もまた
through point (2, negative 4), we can get the same
同じ
answer even when we substitute x equals 2 and y
～するときでも
equals negative 4 into y equals negative $3x$ plus b.

Answer ⑩ y equals negative $3x$ plus 2

52. 1次関数の式を求める問題

問 題 ①y は x の1次関数で、そのグラフは2点$(-1, 5)$, $(2, -4)$を通ります。②この1次関数の式を求めなさい。

解き方 ③求める1次関数の式を $y=ax+b$ とします。④グラフが、2点$(-1, 5)$、$(2, -4)$を通るので、傾き a は、$a=\frac{(-4)-5}{2-(-1)}=-\frac{9}{3}=-3$です。⑤よって、式は $y=-3x+b$ と表わせます。⑥グラフが点$(-1, 5)$を通るので、$x=-1$, $y=5$を $y=-3x+b$ に代入します。⑦$5=-3\times(-1)+b$ より、$b=2$です。⑧したがって、求める式は $y=-3x+2$となります。⑨このグラフは点$(2, -4)$も通るので、$x=2$、$y=-4$を $y=-3x+b$ に代入しても、同じ答えが得られます。

答 え ⑩ $y=-3x+2$

53. Problem of finding an intersection of the graphs of linear functions

Question

① In Figure 1, the equations of straight lines ℓ and m are y equals $3x$ plus 5 and y equals negative x plus 2 respectively.

図　方程式　直線　$y=3x+5$　$y=-x+2$　それぞれ

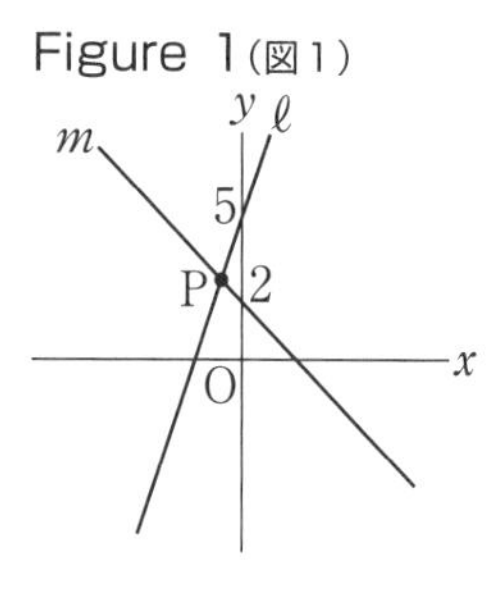

② We suppose P is the intersection point of the 2 straight lines ℓ and m.

考える　交点

③ Find the coordinates of point P.

求めなさい　座標　点

How to solve

④ The coordinates of the intersection point of the 2 straight lines will be found by solving a simultaneous equation of the 2 straight lines.

求められる　解くこと　連立方程式

⑤ The equation of line ℓ is y equals $3x$ plus 5 and the equation of line m is y equals negative x plus 2, therefore, we solve the equations of the 2 by simultaneous equation.

よって　解く

$3x+5=-x+2$ (3x plus 5 equals negative x plus 2)

$4x=-3$ (4x equals negative 3)

$x=-\frac{3}{4}$ (x equals negative 3 over 4)

⑥ In order to find the value of y, substitute x equals negative 3 over 4 into y equals negative x plus 2.

(In order to: ～するために / value: 値 / substitute: 代入する / x equals negative 3 over 4: $x=-\frac{3}{4}$ / into: ～に)

$y=-\left(-\frac{3}{4}\right)+2=\frac{11}{4}$ (y equals negative negative 3 over 4 plus 2 equals 11 over 4)

⑦ As a result, the coordinates of point P becomes (negative 3 over 4, 11 over 4).

(As a result: 結果として)

Answer ⑧ P (negative 3 over 4, 11 over 4)

53. 1次関数のグラフの交点を求める問題

問 題 ①図1で、直線 ℓ、m の式はそれぞれ $y=3x+5$、$y=-x+2$です。②2つの直線 ℓ、m の交点をPとします。③点Pの座標を求めなさい。

解き方 ④2つの直線の交点の座標は、2つの直線の式を連立方程式として解くことで求められます。⑤直線 ℓ の式は $y=3x+5$、直線 m の式は $y=-x+2$だから、2つの式を連立方程式として解きます。
⑥yの値を求めるために、$x=-\frac{3}{4}$を$y=-x+2$に代入します。
⑦したがって、点Pの座標は$\left(-\frac{3}{4}, \frac{11}{4}\right)$となります。

答 え ⑧P$\left(-\frac{3}{4}, \frac{11}{4}\right)$

● 「移動」している最中に……！

54. Use of linear functions (problem of a moving point)

Question

Figure 1（図1）

A　6cm　D
P
4cm
B　C

① Point P moves at a speed of
点　動く　速さ
1 centimeter per second from
秒速1センチメートル(cm)
A through B to C on the
～を通って
perimeter of rectangle ABCD in Figure 1. ② We
周囲　長方形　図
suppose the area of triangle APD x seconds after
考える　面積　△APD　～後
point P starts moving is y square centimeters.
平方センチメートル(cm²)
③ Express y as the expression of x, concerning
表わしなさい　～として　式　～について
when point P moves on each side AB and BC.
～するとき　それぞれの　辺

How to solve

Figure 2（図2）

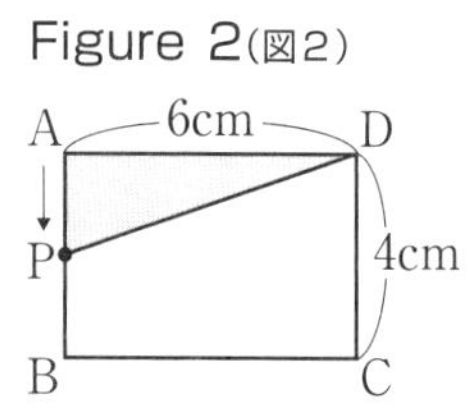

Figure 3（図3）

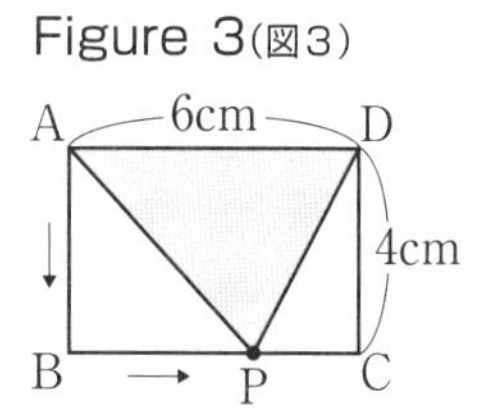

④ When point P moves on side AB, triangle APD
becomes a triangle with base AD and altitude
底辺　高さ

AP.(Figure 2) ⑤Because AP equals x centimeters,
～なので AP＝xセンチメートル
y equals 1 over 2 times 6 times x equals $3x$.
$y=\frac{1}{2}\times 6\times x=3x$
⑥When point P moves on side BC, triangle APD becomes a triangle with base AD and altitude AB. (Figure 3) ⑦So, y equals 1 over 2 times 6 times 4 equals 12.

Answer

⑧Side AB : y equals $3x$, Side BC : y equals 12

54. 1次関数の利用（動点の問題）

問 題 ①図1の長方形ABCDの周上を、点PはAからBを通ってCまで、毎秒1cmの速さで動きます。②点Pが動き始めてからx秒後の△APDの面積を$y$cm^{2}とします。③点Pがそれぞれの辺AB、BC上を動くときについて、yをxの式で表わしなさい。

解き方 ④点Pが辺AB上を動くとき、△APDは底辺AD、高さAPの三角形になります。（図2）⑤AP＝$x$cmだから、$y=\frac{1}{2}\times 6\times x=3x$です。⑥点Pが辺BC上を動くとき、△APDは底辺AD、高さABの三角形になります。（図3）⑦よって、$y=\frac{1}{2}\times 6\times 4=12$です。

答 え ⑧辺AB：$y=3x$、辺BC：$y=12$

●数学の本を投げすてると、「放物線」を描いて……？

55. $y=ax^2$ (introduction)

① When y is the function of x and expressed as y equals
～するとき　関数　表わされる　～として　$y=ax^2$
ax squared, y is said to be in proportion to x squared.
～といわれる　～に比例して
② The graph of the function y equals ax squared is called a parabola, and it has the following
呼ばれる　放物線　次のような～
characteristics.
特徴
③ The graph of the function y equals ax squared passes through the origin and it is a curve line
～を通る　原点　曲線
symmetric to the y axis.
対称な　軸
④ When a is greater than 0, the graph is above the x
$a>0$　～の上側に
axis and will open up.
上に
⑤ On the other hand, when a is less than 0, the
一方　$a<0$
graph is below the x axis and will open down.
～の下側に　下へ
⑥ As the absolute value of a gets larger, the opening
～するにつれて　絶対値　より大きく　開いている
side of the graph becomes smaller. ⑦ In Figure 1, (1)
側　より小さく
is the graph of y equals x squared and (2) is the graph of y equals 1 over 2 times x squared.
$y=\frac{1}{2}x^2$

⑧ In (1) a equals 1, and in (2) a equals 1 over 2, because 1 is greater than 1 over 2, the absolute value of a in (1) is larger, and the opening side of the graph becomes smaller.

（because：～なので）

Figure 1（図1）

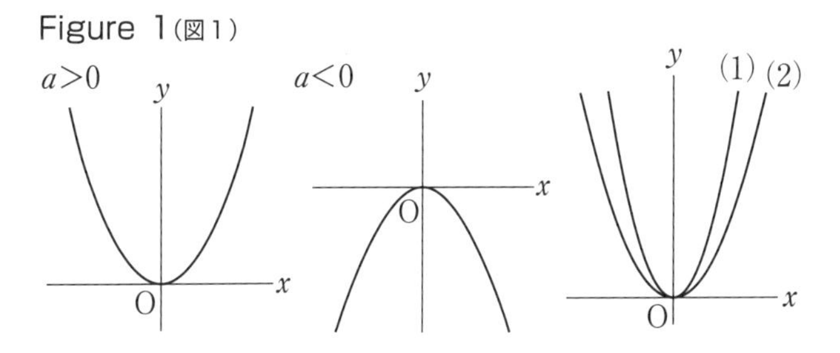

55. $y=ax^2$（導入）

①y が x の関数で、$y=ax^2$で表わされるとき、y は x^2に比例するといいます。②関数 $y=ax^2$のグラフは放物線と呼ばれ、次のような特徴があります。③関数 $y=ax^2$のグラフは、原点を通り、y 軸について対称な曲線です。④ $a>0$のとき、グラフは x 軸の上側にあり、上に開いた形になります。⑤一方、$a<0$のとき、グラフは x 軸の下側にあり、下に開いた形になります。⑥ a の値の絶対値が大きいほど、グラフの開き方は小さくなります。⑦図1で、(1)は $y=x^2$、(2)は $y=\frac{1}{2}x^2$のグラフです。⑧(1)では $a=1$、(2)では $a=\frac{1}{2}$で、$1>\frac{1}{2}$だから、a の値の絶対値が大きい(1)のほうが、グラフの開き方は小さくなります。

● 「変化の割合」が同じなら、答えも……

56. $y=ax^2$ word problem (ratio of change)

Question

① For function y equals x squared, the ratio of change when the value of x increases from a to a plus 3 is equal to the ratio of change of y equals $5x$ minus 7. ② Find the value of a, in this case.

~について 関数 $y=x^2$ 変化の割合 ～するときの 値 増加する $a+3$ 等しい $y=5x-7$ 求めなさい この場合

How to solve

③ In y equals x squared, the value of y when x equals a is y equals a squared. ④ In addition, the value of y when x equals a plus 3 is y equals the square of the quantity a plus 3, close quantity. ⑤ The increased amount of x is the quantity a plus 3, close quantity, minus a equals 3. ⑥ The increased amount of y is the square of the quantity a plus 3, close quantity, minus a squared equals the quantity a squared plus $6a$ plus 9, close quantity, minus a squared equals $6a$ plus 9.

さらに $y=(a+3)^2$ 増加量

⑦ Because the ratio of change equals the increased amount of y over the increased amount of x, it is expressed as $6a$ plus 9 over 3 equals $2a$ plus 3.

～なので　変化の割合$=\frac{y\text{の増加量}}{x\text{の増加量}}$　表わされる　～として

⑧ The ratio of change of y equals $5x$ minus 7 is constant and equal to the coefficient 5 of x. ⑨ We make an equation of a, and solve.

一定な　係数　方程式　解く

$2a+3=5$ (2a plus 3 equals 5)　$2a=2$ (2a equals 2)　$a=1$ (a equals 1)

Answer ⑩ $a=1$ (a equals 1)

56. $y=ax^2$ の文章題（変化の割合）

問　題　①関数 $y=x^2$について、x の値が a から $a+3$まで増加するときの変化の割合が、$y=5x-7$の変化の割合と等しくなります。②このとき、a の値を求めなさい。

解き方　③ $y=x^2$で、$x=a$ のときの y の値は、$y=a^2$です。④また、$x=a+3$のときの y の値は、$y=(a+3)^2$です。⑤ x の増加量は、$(a+3)-a=3$です。⑥ y の増加量は、$(a+3)^2-a^2=(a^2+6a+9)-a^2=6a+9$ です。⑦変化の割合$=\frac{y\text{の増加量}}{x\text{の増加量}}$だから、$\frac{6a+9}{3}=2a+3$と表わせます。⑧ $y=5x-7$の変化の割合は一定で、x の係数5に等しくなります。⑨ a についての方程式をつくり、解きます。

答　え ⑩ $a=1$

57. $y=ax^2$ word problem (problem of finding areas)

Question

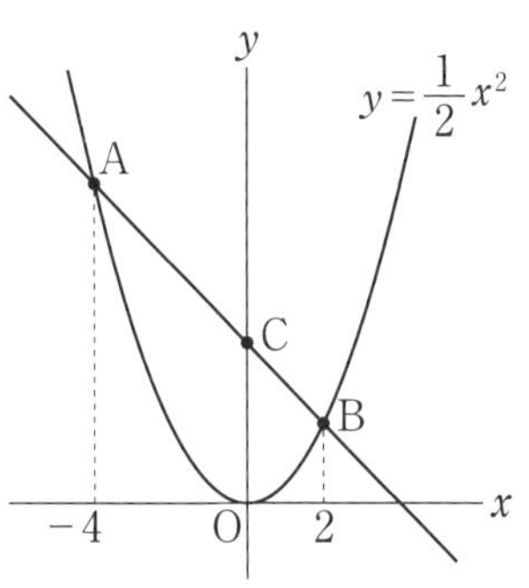

①On the graph of the function y equals 1 over 2 times x squared, there are 2 points A and B.

グラフ / 関数 / $y=\frac{1}{2}x^2$ / 〜がある / 点

② When the x coordinates of A and B are negative 4 and 2 respectively, find the area of triangle OAB.

〜するとき / 座標 / −4 / それぞれ / 求めなさい / 面積 / △OAB

How to solve

③ The y coordinate of point A is (y equals 1 over 2 times negative 4 squared equals) 8 and the y coordinate of point B is (y equals 1 over 2 times 2 squared equals) 2. ④ Because straight line AB passes through 2 points, A(negative 4, 8) and B(2, 2), becomes the graph of y equals negative x plus 4.

$\left(y=\frac{1}{2}\times(-4)^2=\right)8$ / 〜なので / 直線 / 〜を通る / $y=-x+4$

⑤ Thinking the intersection point of straight line AB

考えて / 交点

and the y axis as C. ⑥ The base of triangle OCA and triangle OCB is (OC=) 4, and the altitude becomes the absolute value of the x coordinates of point A and B. ⑦ The area of triangle OCA is (1 over 2 times 4 times 4 equals) 8, that of triangle OCB is (1 over 2 times 4 times 2 equals) 4. ⑧The area of triangle OAB is (8 plus 4 equals) 12.

軸　～として　底辺　高さ　絶対値　(8+4=)12

Answer ⑨12

57. $y=ax^2$ の文章題（面積を求める問題）

問　題　①関数 $y=\frac{1}{2}x^2$のグラフ上に、2点A、Bがあります。②A、Bの x 座標が、それぞれ−4、2であるとき、△OABの面積を求めなさい。

解き方　③点Aの y 座標は $y=\frac{1}{2}\times(-4)^2=8$、点Bの y 座標は $y=\frac{1}{2}\times2^2=2$です。④直線ABは、2点A(−4, 8)、B(2, 2)を通るから、$y=-x+4$のグラフになります。⑤直線ABと y 軸との交点をCとします。⑥△OCA、△OCBは底辺がOC=4で、高さは点A、Bの x 座標の絶対値になります。⑦△OCAの面積は$\frac{1}{2}\times4\times4=8$、△OCBの面積は$\frac{1}{2}\times4\times2=4$です。⑧△OABの面積は、8+4=12です。

答　え　⑨12

Binary scale, decimal scale, hexadecimal scale, sexagesimal scale

2進法，10進法，16進法，60進法

① The numerals that we use every day are 0, 1, 2, 3, 4, 5, 6, 7, 8, 9 and 10, and when it becomes 10, it goes up one position. ② In this way, how to express numbers by using 10 kinds of numerals from 0 to 9 is called the decimal scale. ③ There is also a way to express numbers with only 2 numerals, 0 and 1. ④ This is called the binary scale. ⑤ The binary scale is used in computer data processing. ⑥ On the other hand, in time, 1 minute becomes 60 seconds and 1 hour becomes 60 minutes. ⑦ Like this, the way to express numbers when a position increases by 60 is called sexagesimal scale. ⑧ Also, in programming languages, numbers are expressed with 16 kinds of numerals and alphabets 0, 1, 2, 3, 4, 5, 6, 7, 8, 9, A, B, C, D, E and F. ⑨ This is called the hexadecimal scale. ⑩ In the hexadecimal scale, the next to F becomes 10. ⑪ When we convert the alphabets of hexadecimal scale to decimal scale, A becomes 10, B becomes 11, C becomes 12, D becomes 13, E becomes 14, and F becomes 15.

---

①私たちがふだん使っている数字は、0、1、2、3、4、5、6、7、8、9、10と、10になると1つ位が上がります。②このように0から9までの10種類の数を使って数を表わす方法を、10進法といいます。③0と1の2種類の数字だけで数を表わす方法もあります。④これを2進法といいます。⑤2進法は、コンピュータの情報処理で用いられています。⑥一方、時間は60秒で1分、60分で1時間となります。⑦このように、60で位が上がる数の表わし方を60進法といいます。⑧また、プログラミング言語では、0、1、2、3、4、5、6、7、8、9、A、B、C、D、E、Fの16種類の数字とアルファベットを使って数を表わします。⑨これを16進法といいます。⑩16進法では、Fの次が10になります。⑪16進法のアルファベットを10進法に変換すると、Aは10、Bは11、Cは12、Dは13、Eは14、Fは15になります。

Chapter 4
Figures

第4章　図　形

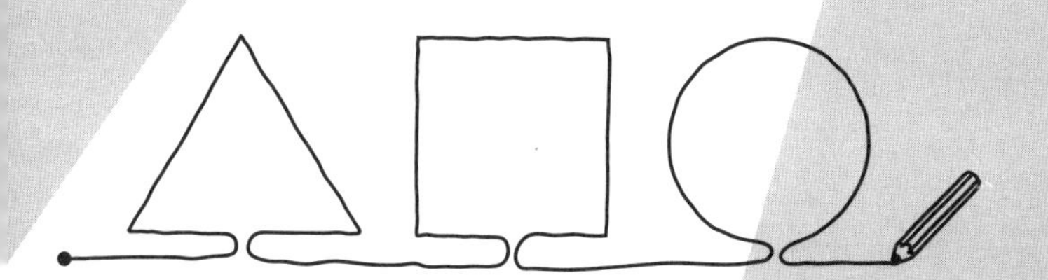

●カドを直角に曲がって、道を「まっすぐ」行くと……

58. Basic vocabulary and concepts of figures

① A line extending straight without end is called a
線 のびている まっすぐに ～なしに 端 呼ばれる
straight line.
直線

② Figure 1 is called straight line AB.
図 直線 AB

③ One part of a straight line with both ends decided
部分 両方の 決められた
is called a line segment.
線分

④ Figure 2 is called line segment AB.
線分 AB

⑤ The length of line segment AB means the distance
長さ 意味する 距離
between point A and point B.
～と…の間の

⑥ The point at the center of the line segment is
真ん中
called the midpoint.
中点

⑦ Point M of Figure 3 is the midpoint of line segment AB.

Figure 1　　Figure 2 　　Figure 3

A B　　A B　　A M B

⑧ When 2 straight lines intersect at right angles,
～するとき 交わる 直角
those 2 lines are perpendicular.
垂直な

⑨ When 2 straight lines intersect vertically, 1 straight line is called a perpendicular line to the other.

vertically: 垂直に　perpendicular line: 垂線　the other: もう一方

58. 図形の基本語句と概念

①限りなくまっすぐにのびる線を直線といいます。

②図1を、直線ABといいます。

③直線の一部で、両方の端が決まっている線を線分といいます。

④図2を、線分ABといいます。

⑤線分ABの長さは、2点A、Bの間の距離ということができます。

⑥線分の真ん中の点を中点といいます。

⑦図3の点Mは、線分ABの中点です。

図1　A　B

図2　A　B

図3　A　M　B

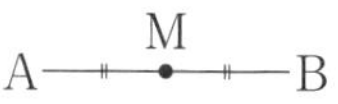

⑧2つの直線が直角に交わるとき、その2つの直線は、垂直であるといいます。

⑨2つの直線が垂直に交わるとき、一方の直線をもう一方の直線の垂線といいます。

●どんな図形も「軸」を決めて回せば、ほら、この通り……

59. Figure transformation

① To（～すること） move the position（位置） of a figure（図形） without（～なしに） changing the size（大きさ） or shape（形） is called（呼ばれる） transformation（移動）. ② In transformation there are（～がある） translation（平行移動）, rotation（回転移動） and reflection（対称移動）. ③ When a（～するとき） figure is moved in a fixed（一定の） direction（方向） and a fixed distance（距離）, it is called translation. ④ In Figure（図） 1, triangle（三角形） DEF is the one that triangle ABC is translated（平行移動された） from.

Figure 1（図1）

A B C D E F

⑤ The transformation of a figure by turning（回転させること） in a fixed angle（角度） around a center（中心） point is called rotation. ⑥ In Figure 2, triangle JKL is a counter-clockwise（反時計回り） 80 degrees（80°） rotation of triangle GHI around center point O.

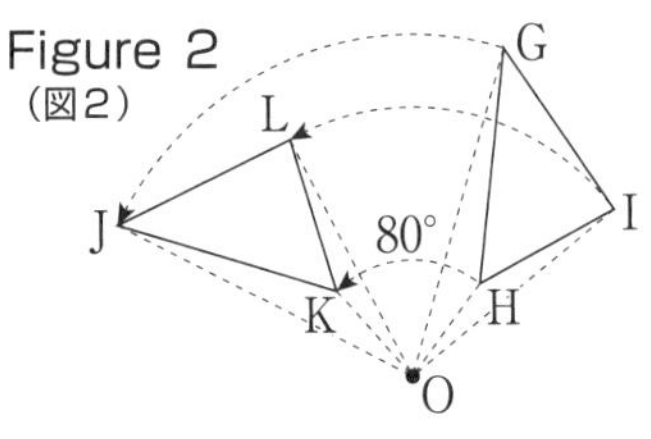

⑦ The transformation of a figure as if folding over a straight line is called reflection.

~かのように 折ること 直線

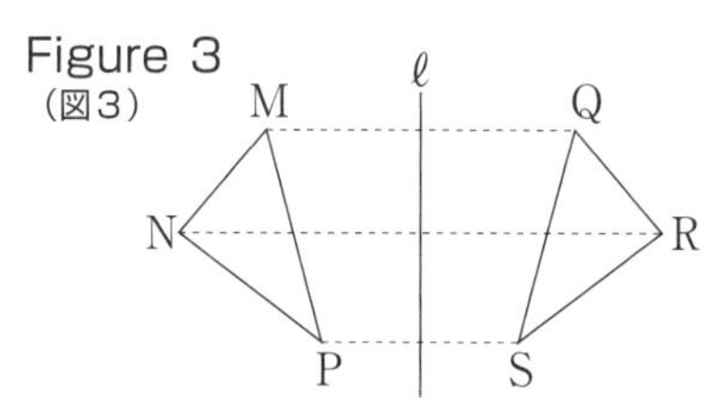

⑧ In Figure 3, triangle QRS is the one that triangle MNP is reflected from thinking straight line ℓ as an axis of reflection.

対称移動された ～として考えて 軸

59. 図形の移動

①図形を、形と大きさを変えずに位置を移すことを移動といいます。②移動には、平行移動、回転移動、対称移動があります。③図形を、一定の方向に、一定の長さだけ動かして移す移動を、平行移動といいます。④図1で、△DEFは、△ABCを平行移動させたものです。⑤図形を、ある1つの点を中心として、一定の角度だけ回転させて移す移動を、回転移動といいます。⑥図2で、△JKLは、△GHIを点Oを中心として、反時計回りに80°回転移動させたものです。⑦図形を、ある直線を折り目として、折り返して移す移動を、対称移動といいます。⑧図3で、△QRSは、△MNPを直線ℓを対称の軸として対称移動させたものです。

●両方の意見は「平行線」のままですが……

60. Parallel lines and angles

Question

① In Figure 1, when straight line ℓ and straight line m are parallel, how many degrees are angle x and angle y respectively?

Figure：図　when：〜するとき　straight line：直線　parallel：平行　how many degrees：何度　angle x：$\angle x$　respectively：それぞれ

Figure 1（図1）

How to solve

② Like in Figure 2, when 1 straight line intersects 2 straight lines, angles in the positions like angle a and angle e are called corresponding angles.

Like：〜のように　intersects：交わる　angles：角　positions：位置　are called：呼ばれる　corresponding angles：同位角

Figure 2（図2）

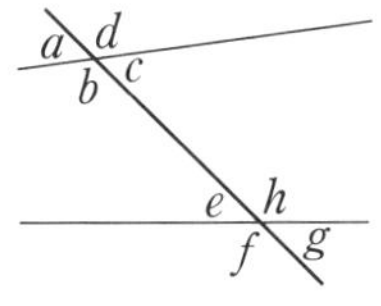

③ Angles in the positions like angle c and angle e are called alternate angles.

alternate angles：錯角

④ When 1 straight line intersects 2 parallel straight lines, the alternate angles will be equal and the

equal：等しい

corresponding angles will be equal, too.

⑤ Therefore, angle x is 75 degrees because it is the
　よって　　　　　　　　75°　　　　～なので
corresponding angle of a 75 degree angle.

⑥ Angle y is 40 degrees because it is the alternate angle of a 40 degree angle.

Answer ⑦ Angle x equals 75 degrees, Angle y equals 40 degrees

60. 平行線と角

問 題 ①図1で、直線 ℓ と直線 m が平行であるとき、$\angle x$、$\angle y$ の大きさはそれぞれ何度ですか。

解き方 ②図2のように、2つの直線に1つの直線が交わるとき、$\angle a$ と $\angle e$ のような位置にある角を同位角といいます。③$\angle c$ と $\angle e$ のような位置にある角を錯角といいます。④平行な2つの直線に1つの直線が交わるとき、錯角は等しく、また、同位角も等しくなります。⑤よって、$\angle x$ は、75°の角の同位角なので、75°です。⑥$\angle y$ は、40°の角の錯角なので、40°です。

答 え ⑦$\angle x = 75°$、$\angle y = 40°$

61. Parallel lines and areas

Question

① In Figure 1, quadrilateral ABCD is a parallelogram and point M is the midpoint of side BC, and also NM and AB are parallel. ② Find all the triangles with an area equal to triangle ABM.

図 四角形 平行四辺形 点 中点 辺 NM// AB 求めなさい 三角形 面積 等しい △ABM

Figure 1（図1）

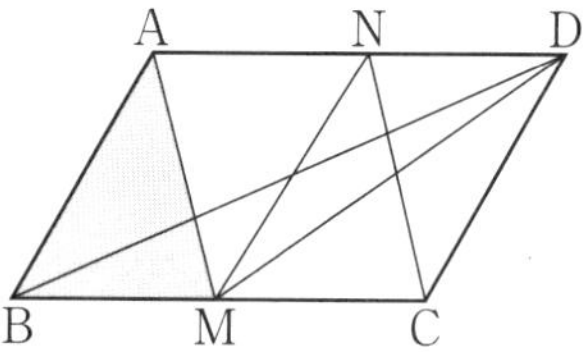

How to solve

③ In Figure 1, in triangle ABM and triangle DBM, if you look at BM as the base of each, because the base is equal and the altitude is equal, the area is equal. ④ If we look at BM and AN as the base of triangle ABM and triangle AMN respectively, because the bases and altitudes are equal, the area of triangle ABM and triangle AMN are equal. ⑤ If we look at BM and MC as the bases of triangle ABM

～なら ～として 底辺 それぞれ ～なので 高さ それぞれ

and triangle NMC respectively, because the bases and altitudes are equal, the area of triangle ABM and triangle NMC are equal. ⑥By the same thinking, we can understand that the areas of triangle DMC and triangle NCD are also equal.

同じ　考え方

Answer ⑦These 6 triangles, triangle DBM, AMN, NMC, DMC, NCD and NMD

61．平行線と面積

問　題 ①図1で、四角形ABCDは平行四辺形で、点MはBCの中点、また、NM// ABです。②△ABMと面積が等しい三角形をすべて答えなさい。

解き方 ③図1で、△ABMと△DBMは、それぞれの底辺をBMとみると、底辺が等しくて高さも等しいので、面積は等しくなります。④△ABMと△AMNの底辺を、それぞれBM、ANとみると、底辺と高さが等しいので、△ABMと△AMNの面積は等しくなります。⑤△ABMと△NMCの底辺を、それぞれBM、MCとみると、底辺と高さが等しいので、△ABMと△NMCの面積は等しくなります。⑥同じように考えると、△DMC、△NCD、△NMDの面積も等しいことがわかります。

答　え ⑦△DBM、△AMN、△NMC、△DMC、△NCD、△NMDの6つ

62. Circles and sectors

Question

① Find the length of the arc and the area of sector OAB like in Figure 1. ② The circular constant is recognized as π.

求めなさい　長さ　弧　面積　おうぎ形　~のような　図　円周率　理解される　~として　パイ

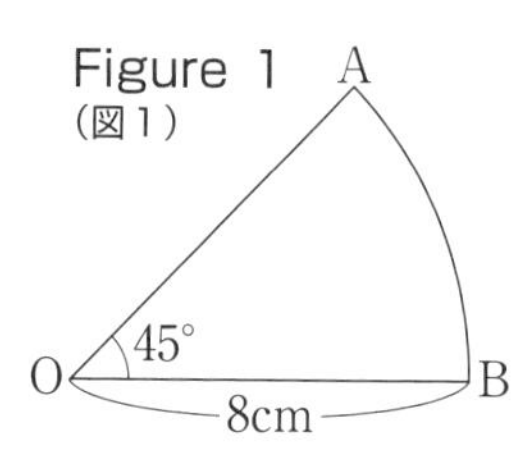

How to solve

③ The circumference of a circle can be found by calculating the diameter (the radius times 2) times the circular constant. ④ The area of a circle can be found by calculating the radius times the radius times the circular constant. ⑤ The length of the arc and the area of a sector is in proportion to the size of the central angle. ⑥ The length of the arc of sector OAB is (2π times 8 times 45 over 360 equals) 2π centimeters. ⑦ The area of a sector with a central angle of 45 degrees is 45 over 360 of the area

円周　円　求められる　計算すること　直径(半径×2)×円周率　半径×半径×円周率　~に比例して　大きさ　中心角

$\left(2\pi\times8\times\frac{45}{360}=\right)2\pi$

センチメートル(cm)　45°

of a circle of the same radius.

⑧ So, the area of sector OAB is (π times 8 squared times 45 over 360 equals) 8π square centimeters.

$\left(\pi \times 8^2 \times \frac{45}{360} = \right) 8\pi$

平方センチメートル(cm^2)

Answer ⑨ length of the arc : 2π centimeters, area : 8π square centimeters

62. 円とおうぎ形

問　題　①図1のようなおうぎ形OABの弧の長さと面積を求めなさい。②円周率はπとします。

解き方　③円の円周は、直径(半径×2)×円周率で求められます。④円の面積は、半径×半径×円周率で求められます。⑤おうぎ形の弧の長さや面積は、中心角の大きさに比例します。⑥おうぎ形OABの弧の長さは、$2\pi \times 8 \times \frac{45}{360} = 2\pi$(cm)です。⑦中心角が45°のおうぎ形の面積は、同じ半径の円の面積の、$\frac{45}{360}$です。⑧よって、おうぎ形OABの面積は、$\pi \times 8^2 \times \frac{45}{360} = 8\pi$($cm^2$)です。

答　え　⑨弧の長さ：2πcm、面積：8πcm^2

● 「内円」の関係は、円満にはおさまりませんね……

63. Inscribed angles

Question

① Find the size of angle x in Figure 1.
求めなさい　大きさ　∠x　図

Figure 1
(図1)

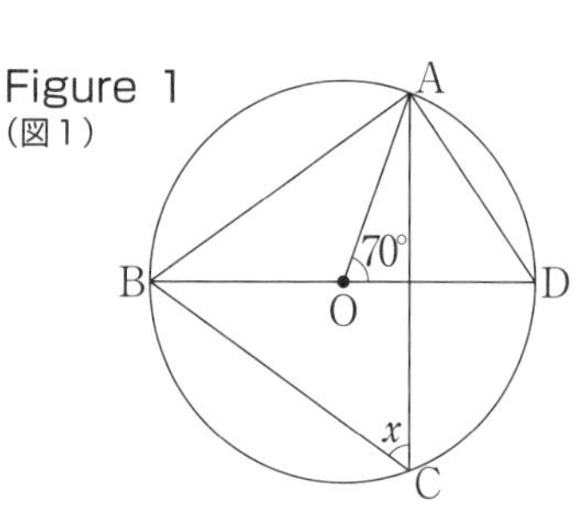

How to solve

② In Figure 1, angle ABD is called an inscribed angle facing arc AD, and angle AOD is called a central angle facing arc AD.
呼ばれる　円周角　向かい合った　弧　中心角

③ The size of the inscribed angle facing an arc is half the size of the central angle facing that arc.
半分

④ So, in Figure 1, the size of angle ABD is (1 over 2 times 70 degrees equals) 35 degrees.
よって

$\left(\frac{1}{2}\times 70^\circ =\right) 35^\circ$

⑤ Because the central angle facing the arc of the semicircle is 180 degrees, the inscribed angle is (1 over 2 times 180 degrees equals) 90 degrees.
〜なので　半円

$\left(\frac{1}{2}\times 180^\circ =\right) 90^\circ$

⑥ So, angle BAD is 90 degrees.

⑦ In triangle ABD, because the sum of the internal angles is 180 degrees, the size of angle ADB is (180 degrees minus 35 degrees minus 90 degrees equals) 55 degrees.

△ABD　和　内角　(180°−35°−90°=) 55°

⑧ Because angle ACB is the inscribed angle facing arc AB, angle x equals angle ADB equals 55 degrees.

Answer ⑨ Angle x equals 55 degrees

63. 円周角

問　題 ①図1で、∠xの大きさを求めなさい。

解き方 ②図1で、∠ABDを弧ADに対する円周角といい、∠AODを弧ADに対する中心角といいます。

③1つの弧に対する円周角の大きさは、その弧に対する中心角の大きさの半分になります。④よって、図1で、∠ABDの大きさは、$\frac{1}{2}\times 70^\circ=35^\circ$です。

⑤半円の弧に対する中心角は180°なので、円周角は$\frac{1}{2}\times 180^\circ=90^\circ$になります。

⑥よって、∠BAD＝90°です。

⑦△ABDで、内角の和は180°なので、∠ADBの大きさは、180°−35°−90°＝55°です。⑧∠ACBは弧ABに対する円周角なので、∠x＝∠ADB＝55°です。

答　え ⑨∠x＝55°

● 「接点」があると、問題が生まれる？

64. Tangent lines of circles

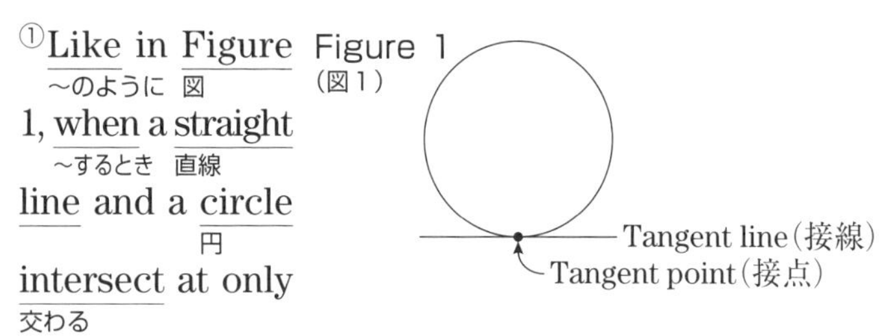

①Like in (～のように) Figure (図) 1, when (～するとき) a straight (直線) line and a circle (円) intersect (交わる) at only 1 point (点), this straight line is said (いわれる) to be tangent (接している) to the circle. ②This straight line is called (呼ばれる) a tangent line (接線) of the circle. ③The point where a circle and a straight line are tangent is called the tangent point (接点).

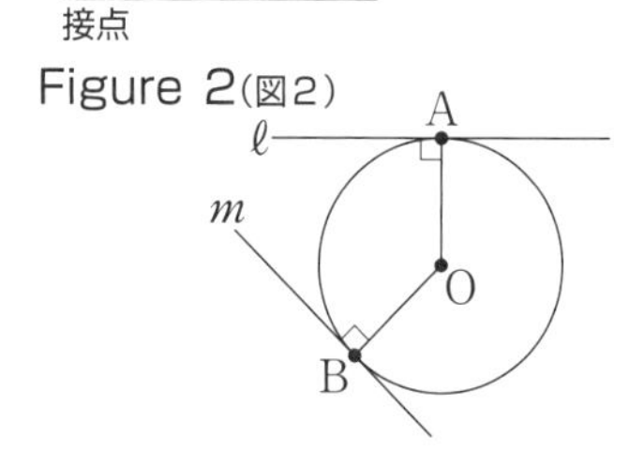

④The tangent line of a circle will be perpendicular (垂直な) to the radius (半径) passing (～を通る) through the tangent point. ⑤In Figure 2, straight line ℓ is perpendicular to straight line OA and straight line m is perpendicular to straight line OB.

⑥Like in Figure 3, there are circle O and point C

outside the circle.
～の外側に

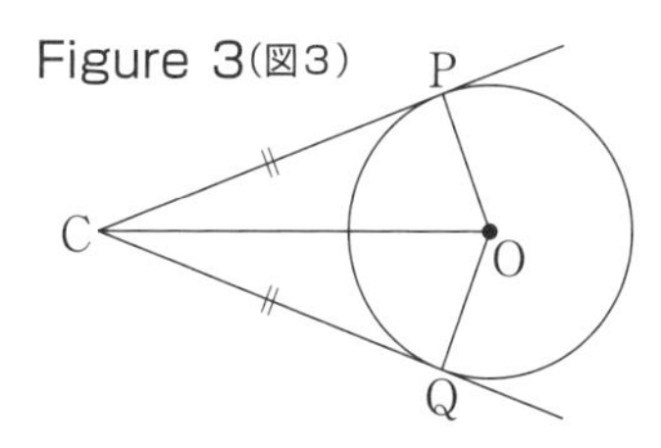

⑦It is possible to draw 2 lines passing through point C which are tangent to circle O. ⑧Line segments CP and CQ of the 2 lines drawn in this case are always equal in length.
可能である　ひく　線分　ひかれた　この場合　必ず　等しい　長さ

⑨The length of 2 tangent lines from a point outside a circle to that circle is always equal.

64. 円の接線

①図１のように、直線と円が１点だけで交わるとき、この直線は円に接するといいます。②この直線を円の接線といいます。③円と直線が接している１点を接点といいます。④円の接線は、接点を通る半径に垂直になります。⑤図２で、直線 ℓ は直線OAに垂直であり、直線 m は直線OBに垂直になっています。⑥図３のように、円Oと円外の点Cがあります。⑦点Cを通る円Oの接線は、２本ひくことができます。⑧このときにできる２本の線分CPとCQは、長さが等しくなります。⑨円外の１点からその円にひいた２本の接線は、必ず長さが等しくなります。

●ある「面」の上にあるもの、ズレているもの……

65. Positional relationship of straight lines and planes

① The positional relationship of straight lines within
位置の 関係 直線 ～内の
a space can be grouped in the following ways.
空間 分類される 次の 方法
② The positional relationship of a straight line and
another straight line can be grouped into 3, when
もう1つの
they intersect like in Figure 1, when they are parallel
交わる ～のように 図 ～するとき 平行な
スキュード
like in Figure 2 and when they are in a skewed position
ねじれの位置
like in Figure 3.

Figure 1（図1） Figure 2（図2）  Figure 3（図3）

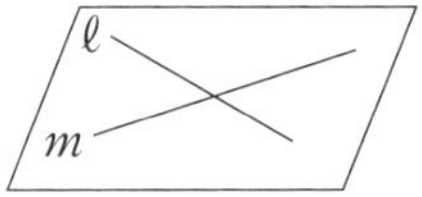

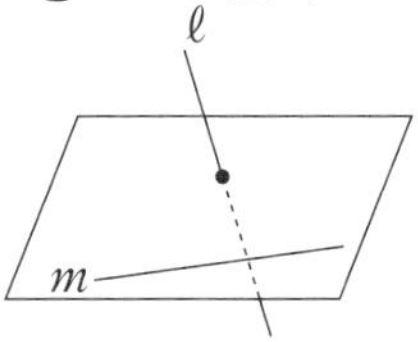

ℓ m ℓ m ℓ m

③ The 2 lines in a skewed position are in a condition
状態
of being neither parallel nor intersecting.
～でも…でもない 交わっている
④ When they intersect and when they are parallel, the
2 lines are on the same plane, but when in a skewed
同じ 平面
position, the two lines are not on the same plane.

⑤ The positional relationship of a straight line and a plane can be grouped into 3, when the straight line is on the plane like in Figure 4, when intersecting like in Figure 5 and when parallel like in Figure 6.

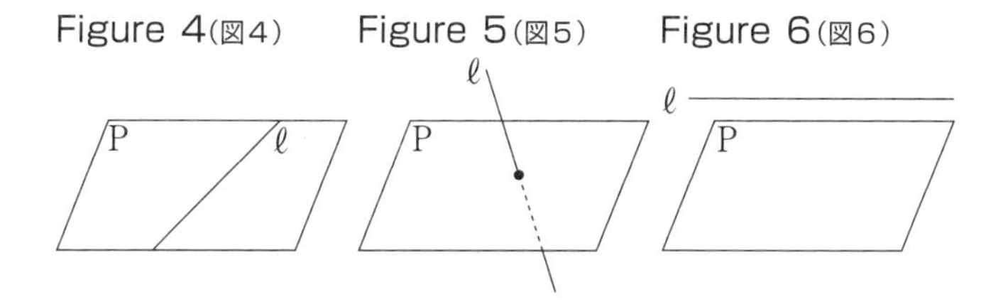

65. 直線と平面の位置関係

①空間内にある直線の位置関係は、次のように分類できます。②直線と直線の位置関係については、図１のように交わる場合、図２のように平行な場合、図３のようにねじれの位置にある場合の３つに分類されます。③ねじれの位置にある２直線は、交わらず、平行でもない状態です。④交わる場合と平行な場合は、２直線は同じ平面上にありますが、ねじれの位置にある場合は、２直線は同じ平面上にありません。⑤直線と平面の位置関係については、図４のように直線が平面上にある場合、図５のように交わる場合、図６のように平行な場合の３つに分類されます。

● 「サッカーボール」って、何面体？

66. Regular polyhedrons

① A solid enclosed by several planes is called a
立体 囲まれた いくつかの 平面 呼ばれる
polyhedron.
多面体
② Depending on the number of planes, polyhedrons
〜によって 数
are called tetrahedron, pentahedron, hexahedron
四面体 五面体 六面体
and so on.
〜など
③ Among polyhedrons, a polyhedron in which every
〜の中で すべての
plane is congruent and the number of planes that
合同
gather at each vertex is always equal and there are
集まる それぞれの 頂点 等しい 〜がない
no dents, it is called a regular polyhedron.
へこみ 正多面体
④ There are only 5 kinds of regular polyhedron,
種類
regular tetrahedron, regular hexahedron, regular
正四面体 正六面体 正八面体
octahedron, regular dodecahedron, and regular
正十二面体 正二十面体
icosahedron.

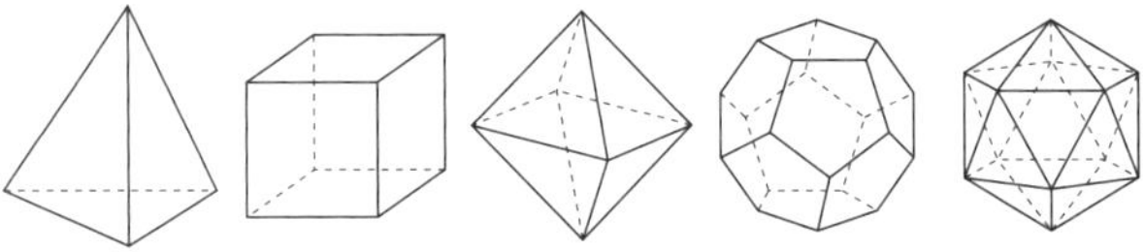

Regular tetrahedron | Regular hexahedron | Regular octahedron | Regular dodecahedron | Regular icosahedron

⑤ A regular hexahedron is enclosed by only 6 congruent squares.
(squares：正方形)

⑥ A regular hexahedron and a cube are identified as the same solid.
(cube：立方体　are identified：定義される　as：〜として)

66. 正多面体

①いくつかの平面で囲まれている立体を多面体といいます。
②多面体は、その面の数によって、四面体、五面体、六面体などといいます。
③多面体の中でも、すべての面が合同である正多角形で、各頂点に集まる面の個数がすべて等しく、へこみのないものを正多面体といいます。
④正多面体には、正四面体、正六面体、正八面体、正十二面体、正二十面体の5種類しかありません。

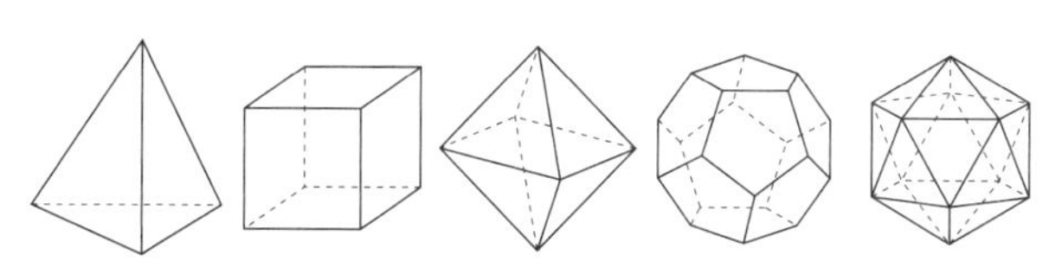

正四面体　正六面体　正八面体　正十二面体　正二十面体

⑤正六面体は6つの合同な正方形だけで囲まれています。
⑥正六面体と立方体は同じ立体を指しています。

● 「影」のある人って、何かがある?!

67. Projection drawing

① As a way to express a solid on a plane, there is the

～として 方法 表わす 立体 平面 ～がある

projection drawing.

投影図

② Like in Figure 1, let's think

～のように 図

about when parallel light

～するとき 平行な 光線

rays hit a triangular prism.

あてる 三角柱

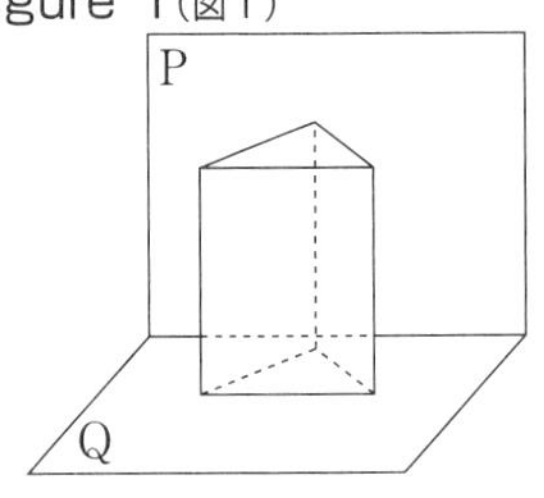

Figure 1（図1）

③ The shadow cast by the

影 (影を)投げかけられた

light rays perpendicular to

垂直な

plane P is a rectangle. ④ The shadow cast by the

長方形

light rays perpendicular to plane Q is a triangle.

三角形

⑤ The shadow cast on plane P is the figure seen when

図形 見える

viewed from directly in front of the triangular prism.

見られた ちょうど ～の正面

⑥ This figure seen from directly in front is called the

呼ばれる

elevation view.

立面図

⑦ The shadow cast on plane Q is the figure seen

when viewed from directly above the triangular

～の上方

prism.

⑧This figure seen from directly above is called the plan view.⑨A projection drawing is expressed, like in Figure 2, as a combination of this elevation view and plan view.

平面図 / 表わされる / 組み合わせ

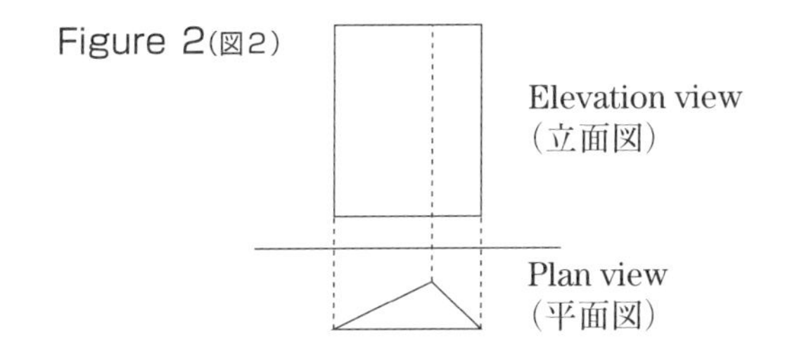

67. 投影図

①立体を平面上に表わす方法として、投影図があります。②図1のように、三角柱に平行な光線をあてた場合で考えてみましょう。③平面Pに垂直な光線によってできる影は、長方形になります。④平面Qに垂直な光線によってできる影は、三角形になります。⑤平面Pにできる影は、三角柱を真正面から見たときに見える図形です。⑥この真正面から見た図を立面図といいます。⑦平面Qにできる影は、三角柱を真上から見たときに見える図形です。⑧この真上から見た図を平面図といいます。⑨投影図は、この立面図と平面図を組み合わせて、図2のように表わします。

●「三角帽子」って、やっぱりかぶりにくい？

68. Volume of conic solids

Question

① Find the volume of the following solids.
求めなさい 体積 次の 立体

② The circular constant is recognized as π
円周率 理解される ～として

(1)

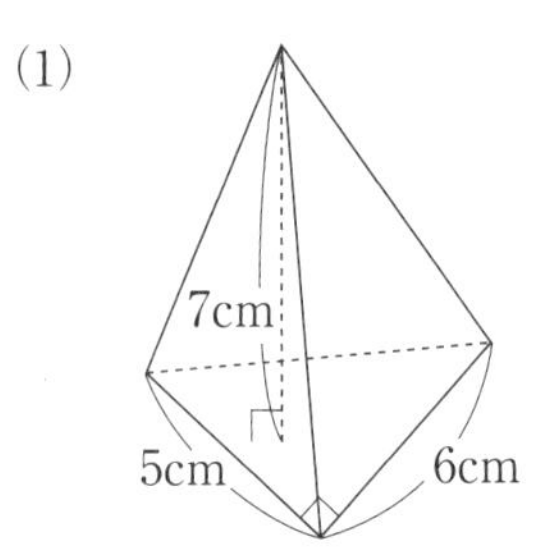

(2)

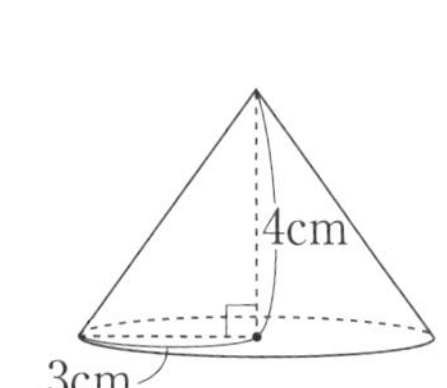

How to solve

③ (1) is called a triangular pyramid. ④ (2) is called a cone. ⑤ The volume of a conic solid can be found by multiplying 1 over 3 times the base area times the altitude. ⑥ The base area of the triangular pyramid in (1) is (1 over 2 times 5 times 6 equals) 15 square centimeters. ⑦ Because the altitude is 7

三角錐 円錐 求められる かけること $\frac{1}{3}$×底面積×高さ

$\left(\frac{1}{2}\times5\times6=\right)15$ 平方センチメートル(cm^2) ～なので

centimeters, the volume is (1 over 3 times 15 times 7 equals) 35 cubic centimeters. ⑧The base area of the cone in (2) is (π times 3 squared equals) 9π square centimeters. ⑨Because the altitude is 4 centimeters, the volume is (1 over 3 times 9π times 4 equals) 12π cubic centimeters.

cubic centimeters: 立方センチメートル(cm^3)
π times 3 squared equals 9π: $\pi \times 3^2 = 9\pi$

Answer ⑩(1) 35 cubic centimeters, (2) 12π cubic centimeters

68. 錐体の体積

問　題　①次の立体の体積を求めなさい。②円周率はπとします。

解き方　③(1)は三角錐(すい)といいます。④(2)は円錐といいます。⑤錐体の体積は、$\frac{1}{3}\times$底面積$\times$高さで求められます。⑥(1)の三角錐の底面積は、$\frac{1}{2}\times5\times6=15(cm^2)$です。⑦高さは7 cmなので、体積は、$\frac{1}{3}\times15\times7=35(cm^3)$です。⑧(2)の円錐の底面積は、$\pi\times3^2=9\pi\ (cm^2)$です。⑨高さは4 cmなので、体積は、$\frac{1}{3}\times9\pi\times4=12\pi\ (cm^3)$です。

答　え　⑩(1) $35cm^3$　(2) $12\pi\ cm^3$

●「展開図」さえ書ければ、解けたも同然……？

69. Surface areas of cones

Question

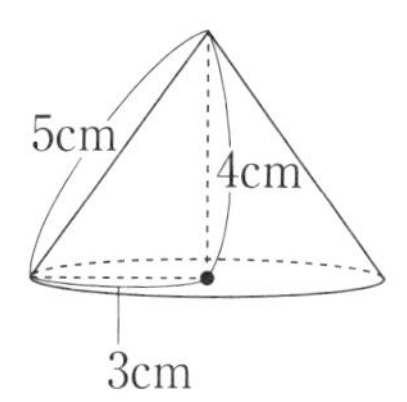

①Find the surface area of the following cone.
求めなさい　表面積　次の～　円錐

②The circular constant is recognized as π.
円周率　理解される　～として

How to solve

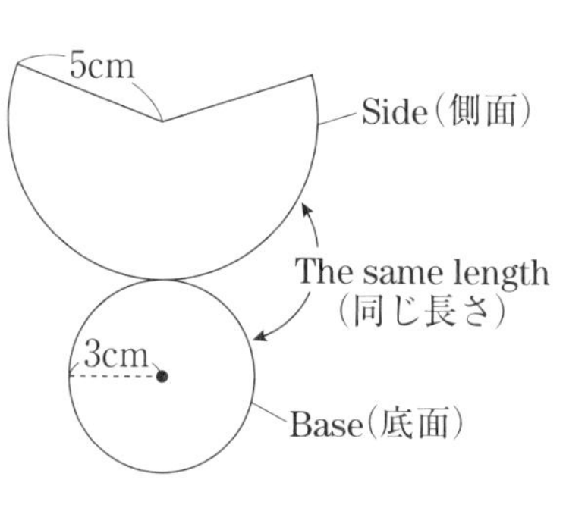

③The surface area of a cone can be found by calculating the base area plus the lateral area.
求められる　計算すること　底面積＋側面積

④The base area is (π times 3 squared equals) 9π square centimeters.
$(\pi\times3^2=)9\pi$　平方センチメートル(cm^2)

⑤In the development view, the side is the sector with a radius of 5 centimeters.
展開図　側面　おうぎ形　半径

⑥As shown in the development view, the length of the base circle's circumference is equal to the length of the arc of the sector's side development.
～ように　示された　長さ　底面　円　円周　等しい　弧

⑦The

side of this solid is a sector with a size (2π times 3 over 2π times 5 equals) 3 over 5 of a circle with a 5 centimeter radius. ⑧The lateral area is (π times 5 squared times 3 over 5 equals) 15π square centimeters. ⑨The surface area is (9π plus 15π equals) 24π square centimeters.

solid 立体　size 大きさ　$\left(\frac{2\pi\times3}{2\pi\times5}=\right)\frac{3}{5}$　$\left(\pi\times5^2\times\frac{3}{5}=\right)15\pi$　$(9\pi+15\pi=)24\pi$

Answer ⑩24π square centimeters

69. 円錐の表面積

問　題 ①次の円錐の表面積を求めなさい。②円周率はπとします。

解き方 ③円錐の表面積は、底面積＋側面積で求められます。④底面積は、$\pi\times3^2=9\pi$(cm²)です。⑤展開図で、側面は、半径が5cmのおうぎ形になります。⑥展開図で示したように、底面の円の円周の長さと、側面の展開図のおうぎ形の弧の長さは等しくなります。⑦この立体の側面は、半径が5cmの円の、$\frac{2\pi\times3}{2\pi\times5}=\frac{3}{5}$の大きさのおうぎ形です。⑧側面積は、$\pi\times5^2\times\frac{3}{5}=15\pi$(cm²)です。⑨表面積は、$9\pi+15\pi=24\pi$(cm²)です。

答　え ⑩24π(cm²)

●どこを「軸」にするかが問題

70. Solids of revolution

Question

① In the following figures, find the volume of each
次の　図　求めなさい　体積　それぞれの
solid made when (1) and (2) are rotated 1 turn in
立体　つくられた　～するとき　回転された　回転
the direction of the arrow.
方向　矢印
② The circular constant is recognized as π.
円周率　理解される　～として

(1)

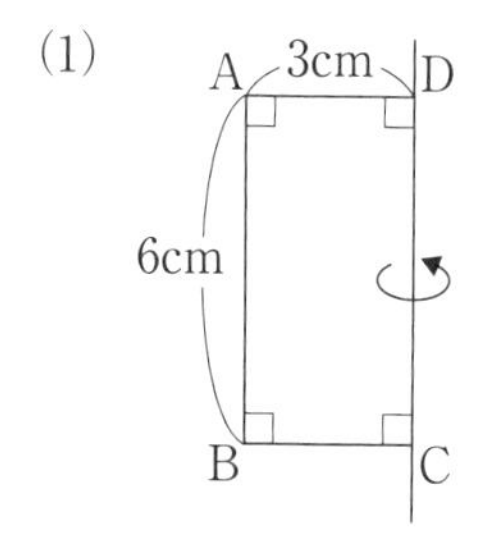

(2)

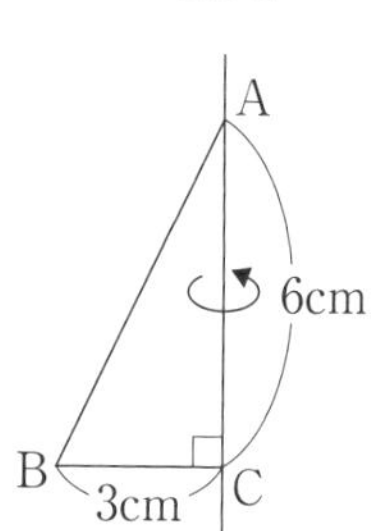

How to solve

③ The solid made when a figure is rotated around a
straight line as an axis is called a solid of revolution.
直線　軸　呼ばれる　回転体
④ (1) becomes a cylinder. ⑤ (2) becomes a cone.
円柱　円錐
⑥ The volume of the cylinder is, from π times the
$\pi \times$(半径)$^2\times$高さ
radius squared times altitude, (π times 3 squared
($\pi \times 3^2 \times 6 =$)$54\pi$

times 6 equals) 54 π cubic centimeters.
立方センチメートル(cm³)

⑦ The volume of the cone is, from 1 over 3 times π
$\frac{1}{3}\times\pi\times$(半径)$^2\times$高さ
times radius squared times altitude, (1 over 3 times π times 3 squared times 6 equals) 18π cubic centimeters.

Answer ⑧(1) 54 π cubic centimeters (2) 18π cubic centimeters

70. 回転体

問 題 ①次の図で、(1)、(2)を矢印の向きに1回転させたときにできる立体の体積をそれぞれ求めなさい。②円周率はπとします。

解き方 ③1つの直線を軸として回転させたときにできる立体を回転体といいます。
④(1)は円柱になります。⑤(2)は円錐になります。
⑥円柱の体積は、$\pi\times$(半径)$^2\times$高さで、$\pi\times3^2\times6=54\pi$ (cm³)です。
⑦円錐の体積は、$\frac{1}{3}\times\pi\times$(半径)$^2\times$高さで、$\frac{1}{3}\times\pi\times3^2\times6=18\pi$ (cm³)です。

答 え ⑧(1) 54π cm³ (2) 18π cm³

●「合同」って、まったく同じってこと？

71. Congruent conditions of triangles

① Like triangle ABC and triangle DEF in Figure 1,
～のような △ABC 図
two figures which lie on top of each other exactly
重なる ぴったり
are called congruent figures. ② When we show that
～と呼ばれる 合同な ～するとき 示す
triangle ABC and DEF are congruent, it is expressed
表わされる
as $\triangle ABC \equiv \triangle DEF$ (triangle ABC is congruent with triangle DEF).
～として

Figure 1

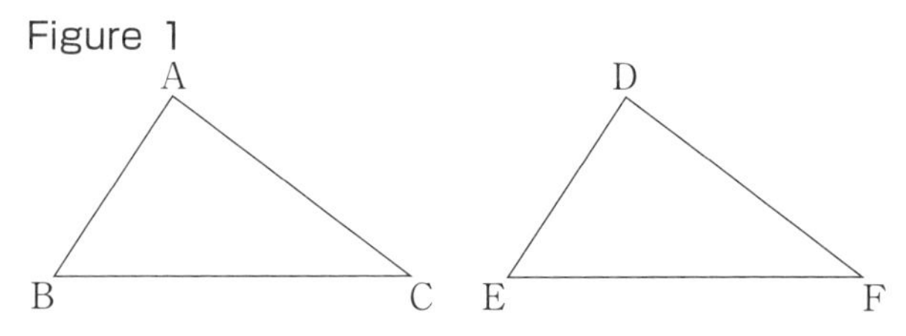

③ The congruent conditions of triangles are the
合同条件
following 3. ④ The first is, "3 sets of the sides are
次の 1つ目 3組 辺
each equal." ⑤ The second is, "2 sets of the sides
それぞれ 等しい 2つ目
and the angle between them are each equal." ⑥ If
角 ～の間の ～なら
the position of the angle whose size is determined is
位置 大きさ 決められた
between 2 sides, 1 triangle is identified.
特定される
⑦ The third is, "a set of the sides and the angles at
3つ目

both ends are each equal."
両方の 端

⑧If the position of the angles whose sizes are determined is at both ends of 1 side, 1 triangle is identified.

71．三角形の合同条件

①図1の△ABCと△DEFのように、ぴったりと重なる2つの図形を合同な図形といいます。②△ABCと△DEFが合同であることを示すときは、△ABC≡△DEFと表わします。

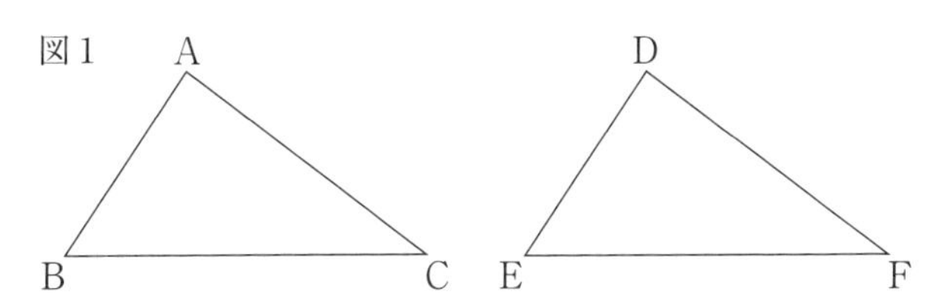

③三角形の合同条件には、次の3つがあります。④1つ目は、「3組の辺がそれぞれ等しい」です。⑤2つ目は、「2組の辺とその間の角がそれぞれ等しい」です。⑥大きさが決まっている角の位置が、2辺の間であれば、三角形は1つに決まります。⑦3つ目は、「1組の辺とその両端の角がそれぞれ等しい」です。⑧大きさが決まっている角の位置が、1辺の両端であるときだけ、三角形は1つに決まります。

●四角いものは、「変形」しやすい？

72. Special parallelograms

① A quadrilateral with 2 sets of facing sides which
四角形 2組 向かい合う 辺
are each parallel is called a parallelogram.
それぞれ 平行な 呼ばれる 平行四辺形

② Because a rectangle, a rhombus and a square also
～なので 長方形 ひし形 正方形 ～もまた
have 2 sets of facing sides which are each parallel,
they can be called special shapes of parallelogram.
特別な 形

③ When we add the condition "the angles next to
～するとき 加える 条件 角 ～のとなりの
each other are equal" to a parallelogram, it becomes
お互い 等しい
a rectangle.

④ When we add the condition "the sides next to each other are equal" to a parallelogram, it becomes a rhombus.

⑤ When we add the condition "the sides next to each other are equal" to a rectangle, it becomes a square.

⑥ When we add the condition "the angles next to each other are equal" to a rhombus, it also becomes a square.

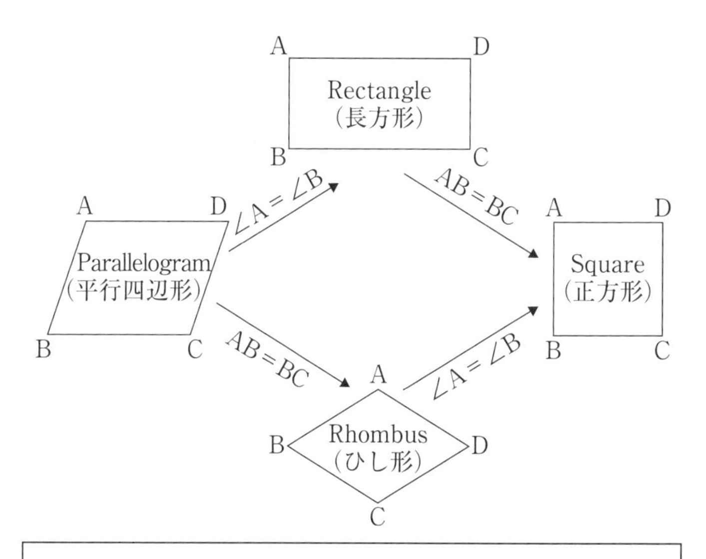

72. 特別な平行四辺形

①２組の向かい合う辺がそれぞれ平行である四角形を平行四辺形といいます。

②長方形、ひし形、正方形も、２組の向かい合う辺がそれぞれ平行なので、平行四辺形の特別な形といえます。

③平行四辺形に「となり合う角が等しい」という条件を加えると長方形になります。

④平行四辺形に「となり合う辺が等しい」という条件を加えるとひし形になります。

⑤長方形に「となり合う辺が等しい」という条件を加えると正方形になります。

⑥ひし形に「となり合う角が等しい」という条件を加えても正方形になります。

● 「証明」するには、条件が満たされないと……

73. Proof of congruence

Question

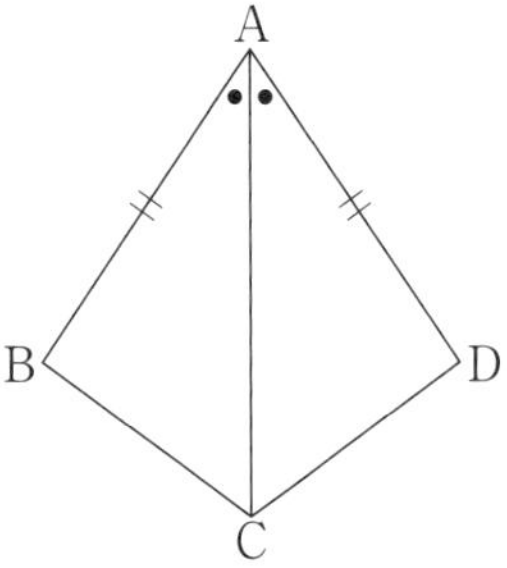

①In quadrilateral ABCD, the lengths of AB and AD are equal and the size of angle BAC and angle DAC are equal.

四角形 / 長さ / 等しい / 大きさ / ∠BAC

②In this case, prove that triangle ABC and triangle ADC are congruent.

この場合 / 証明しなさい / △ABC≡△ADC

How to solve

③First, we look for angles with the same size and sides with the same length in the 2 triangles to prove the congruence.

まず / 探す / 同じ

④Next, we think which congruent condition can be applied.

次に / 合同条件 / あてはめられる

⑤Finally, we show the congruent condition and prove the congruence.

最後に

Answer ⑥As for triangles ABC and ADC,

〜についていえば

[7] From the assumption, AB equals AD…(1)
仮定　AB = AD

[8] From the assumption, angle BAC equals angle DAC…(2)

[9] Because it is common, AC equals AC…(3)
～なので　共通

[10] From (1), (2) and (3), triangle ABC and triangle ADC are congruent because 2 sets of sides and the angle between them are each equal.
2組　辺　～の間の　それぞれ

73. 合同の証明

問　題　[1]四角形 ABCD で、AB と AD は長さが等しく、∠BAC と∠DAC は大きさが等しくなっています。[2]このとき、△ABC≡△ADC であることを証明しなさい。

解き方　[3]まず、証明をするために、同じ大きさの角度や、同じ長さの辺を２つの三角形から探します。[4]次に、どの合同条件があてはまるかを考えます。[5]最後に、合同条件を示して、合同を証明します。

答　え　[6]△ABC と△ADC において、
[7]仮定より、AB＝AD…(1)
[8]仮定より、∠BAC ＝∠DAC…(2)
[9]共通なので、AC＝AC…(3)
[10](1)、(2)、(3)より、２組の辺とその間の角がそれぞれ等しいので、△ABC≡△ADC

●「相似」って、似たもの同士のこと？

74. Similarity conditions of triangles

① In Figure 1, triangle DEF is a figure which triangle
図 △DEF
ABC is reduced. ② Figures like this are called
縮小された ～のような 呼ばれる
similar figures. ③ When we show that triangle ABC
相似な ～するとき 示す
and triangle DEF are similar, it is expressed as
表わされる ～として
△ABC∽△DEF (triangle ABC is similar to triangle DEF).

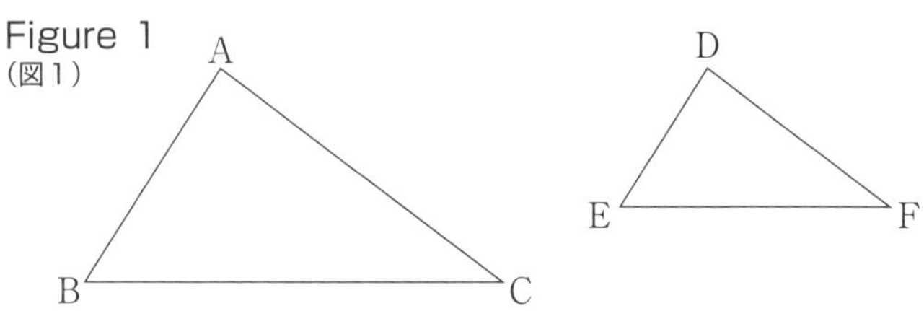

④ There are the following 3 similarity conditions to
～がある 次の 相似条件
show that 2 triangles are similar. ⑤ The first is, "the
1つ目
proportions of the 3 sets of the sides are all equal."
比 3組 辺 すべて 等しい
⑥ The second is, "the proportion of the 2 sets of the
2つ目
sides and the angle between them are each equal."
角 ～の間の それぞれ
⑦ The third is, "the 2 sets of the angles are each equal."
3つ目
⑧ In triangles, if the size of 2 angles is determined,
～なら 大きさ 決められる

the size of the remaining angle is determined.
残りの
⑨ Triangles with three equal angles become the same shape triangles. ⑩ Therefore, we can say that when the size of 2 angles are each equal, the triangles are similar.
同じ 形 よって

74. 三角形の相似条件

①図1で、△DEFは△ABCを縮小した形になっています。
②このような図形を相似な図形といいます。
③△ABCと△DEFが相似であることを示すときは、△ABC∽△DEFと表わします。
④2つの三角形が相似であることを示すための相似条件には、次の3つがあります。
⑤1つ目は、「3組の辺の比がすべて等しい」です。
⑥2つ目は、「2組の辺の比とその間の角がそれぞれ等しい」です。
⑦3つ目は、「2組の角がそれぞれ等しい」です。
⑧三角形は2つの角の大きさが決まれば、残りの1つの角の大きさが決まります。
⑨3つの角が等しい三角形は、形が同じ三角形になります。
⑩したがって、2つの角の大きさがそれぞれ等しい三角形は、相似であるといえます。

75. Proof of similarity

Question

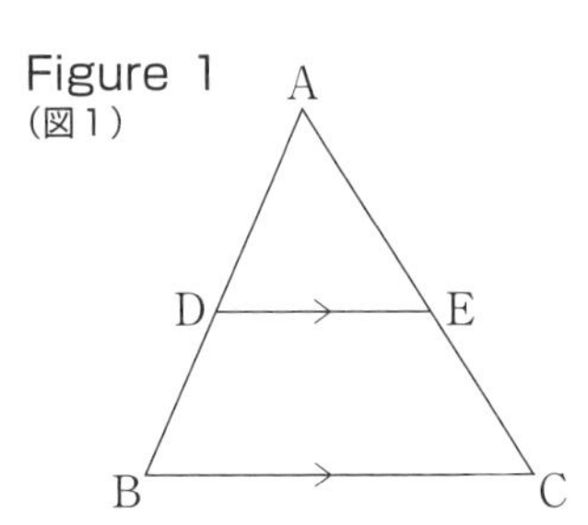

① In triangle ABC, point D is a point on side AB and point E is a point on side AC.
△ABC 点 辺

② When DE and BC are parallel, prove that triangle ABC and triangle ADE are similar.
～するとき DE// BC 証明しなさい
△ABC∽△ADE

How to solve

③ From DE and BC being parallel, because the corresponding angles are equal, angle ABC equals angle ADE. ④ Because angle A is common, angle BAC equals angle DAE. ⑤ Because 2 sets of angles are equal, we can say triangle ABC and triangle ADE are similar. ⑥ In the proof of similarity, it is enough to show that 2 sets of angles are equal.
～なので 同位角 等しい ∠ABC＝∠ADE 共通 2組 証明 相似 十分 示す

Answer ⑦ As for triangle ABC and triangle ADE,
〜についていえば
⑧ From DE and BC being parallel, because the corresponding angles are equal, angle ABC equals angle ADE…(1) ⑨ Because angle A is common, angle BAC equals angle DAE…(2) ⑩ From (1) and (2), because 2 sets of angles are each equal, ⑪ Triangle ABC and triangle ADE are similar.
それぞれ

75. 相似の証明

問　題　①△ABC で、点 D は辺 AB 上の点で、点 E は辺 AC 上の点です。②DE// BC であるとき、△ABC∽△ADE であることを証明しなさい。

解き方　③DE// BC より、同位角が等しいので、∠ABC＝∠ADE です。④∠A は共通なので、∠BAC＝∠DAE です。⑤２組の角が等しいので、△ABC∽△ADE といえます。⑥相似の証明では、２組の角が等しいことを示せばよいです。

答　え　⑦△ABC と△ ADE において、
⑧DE// BCより同位角が等しいので、∠ABC＝∠ADE…(1)
⑨∠Aは共通なので、∠BAC＝∠DAE…(2)
⑩(1)、(2)より、２組の角がそれぞれ等しいので、
⑪△ABC∽△ADE

●「中間点」まで同じだと……

76. Midpoint theorem

Question

① In trapezoid ABCD that AD
台形
and BC are parallel, point M
AD // BC 点
is the midpoint of side AB.
中点 辺
② We express the intersect
表わす 交点
point of side CD and a straight line parallel to side
直線
AD and passing through point M as N. ③ How many
~を通る ~として 何センチメートル(cm)
centimeters is the length of line segment MN?
長さ 線分

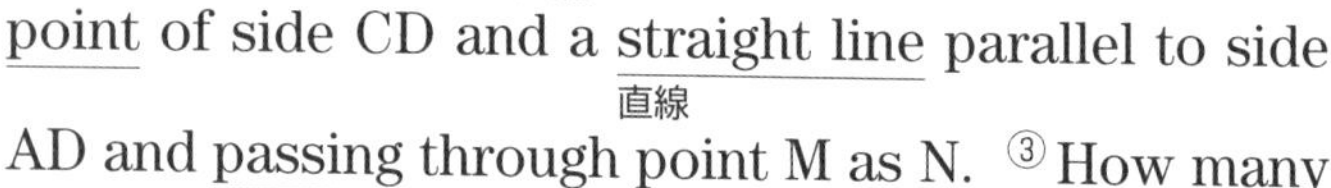

How to solve

④ In Figure 1, when we express the midpoints of the
図 ~するとき
2 sides AB and AC of triangle ABC as point M and
△ABC
point N respectively, MN and BC are parallel and
それぞれ
MN equals 1 over 2 times BC. ⑤ This is called the
$MN=\frac{1}{2}BC$ 呼ばれる
midpoint theorem. ⑥ In triangle ABC of Figure 2,
中点連結定理
from the midpoint theorem, ME equals 1 over 2

times BC equals 1 over 2 times 6 equals 3 centimeters.
センチメートル
⑦ In triangle ACD, from the midpoint theorem, EN equals 1 over 2 times AD equals 1 over 2 times 4 equals 2 centimeters. ⑧ MN equals ME plus EN equals 3 plus 2 equals 5 centimeters.
MN=ME+EN=3+2=5

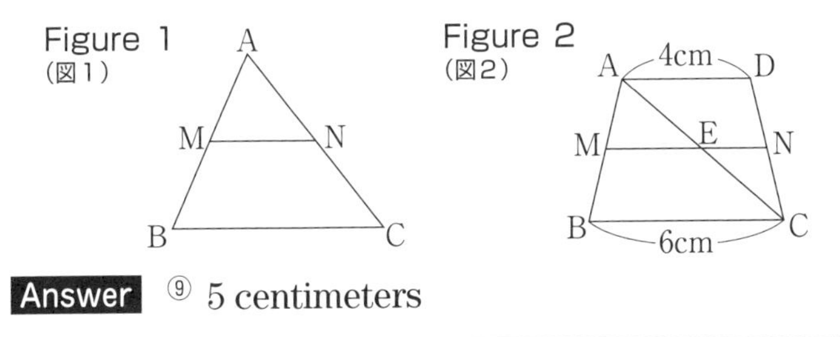

Answer ⑨ 5 centimeters

76. 中点連結定理

問　題　①AD// BCである台形ABCDで、点Mは辺ABの中点です。②点Mを通り、辺ADに平行な直線と辺CDとの交点をNとします。③線分MNの長さは何cmですか。

解き方　④図1で、△ABCの2辺AB、ACの中点をそれぞれ点M、点Nとすると、MN// BCであり、$MN=\frac{1}{2}BC$になります。⑤これを、中点連結定理といいます。⑥図2の△ABCで、中点連結定理より、$ME=\frac{1}{2}BC=\frac{1}{2}\times 6=3$(cm)です。⑦△ACDで、中点連結定理より、$EN=\frac{1}{2}AD=\frac{1}{2}\times 4=2$(cm)です。⑧MN＝ME＋EN＝3＋2＝5(cm)です。

答　え　⑨5cm

●三角形の「斜辺」にも言い分はある？

77. The Pythagorean theorem

Question

① Triangle ABC is a right triangle with angle ACB being 90 degrees.
△ABC 直角三角形 ∠ACB 90°

② Find the length of side AB.
求めなさい 長さ 辺

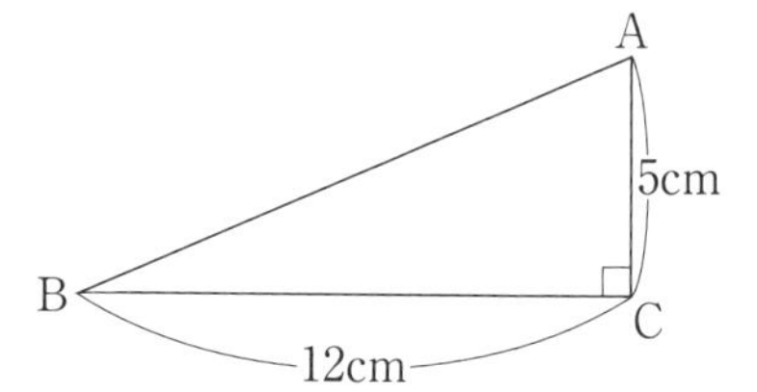

How to solve

③ A right triangle like in Figure 1 is made up of the relationship a squared plus b squared equals c squared.
〜のような 図 成り立つ 関係 $a^2+b^2=c^2$

Figure 1（図1）

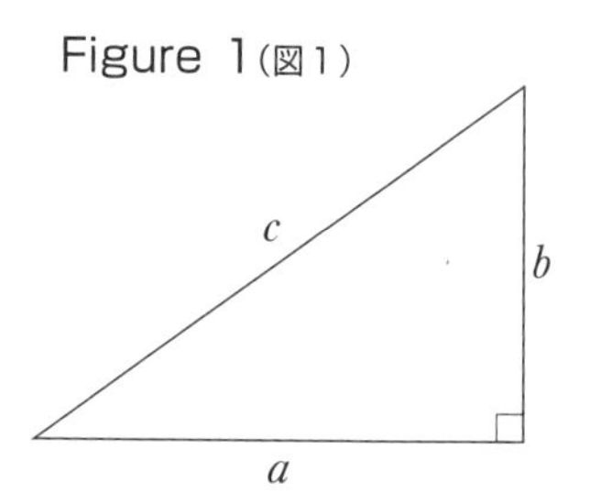

④ This is called the Pythagorean theorem.
呼ばれる 三平方の定理

⑤ The lengths of the 2 sides sandwiching the right angle of right triangle ABC are BC equals 12 centimeters and AC equals 5 centimeters.
はさんでいる 直角 BC＝12センチメートル(cm)

⑥So, AB squared equals BC squared plus AC squared equals 12 squared plus 5 squared equals 144 plus 25 equals 169.
$AB^2=BC^2+AC^2=12^2+5^2=144+25=169$

⑦AB equals positive or negative 13.
AB＝±13

⑧Because the length of AB is a positive number, AB equals 13 centimeters.
Because：～なので　positive：正の　number：数

Answer ⑨13 centimeters

77. 三平方の定理

問　題　①△ABCは、∠ACB＝90°の直角三角形です。
②辺ABの長さを求めなさい。

解き方　③図1のような直角三角形では、$a^2+b^2=c^2$という関係が成り立ちます。
④これを、三平方の定理といいます。
⑤直角三角形ABCの直角をはさむ2辺の長さは、BC＝12cm、AC＝5cmです。
⑥よって、$AB^2=BC^2+AC^2=12^2+5^2=144+25=169$
⑦AB＝±13となります。
⑧ABの長さは正の数なので、AB＝13cmです。

答　え　⑨13cm

78. The Pythagorean theorem (Altitudes of regular triangles)

Question

① Find the area of a regular triangle with a side of 6 centimeters.
求めなさい　面積　正三角形　辺
センチメートル(cm)

How to solve

② Like in Figure 1, vertices A,B and C are decided.
～のように　図　頂点(複数)　定められる

Figure 1
(図1)

③ We draw a vertical line of side BC passing through vertex A, and D is the intersection point of side BC.
ひく　垂線　～を通っている　頂点(単数)　交点

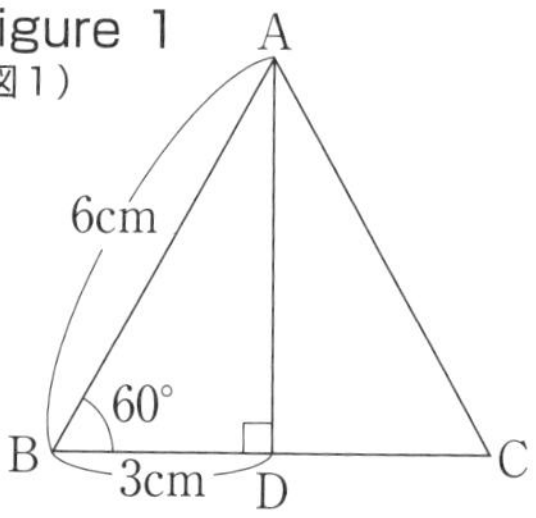

④ Triangle ABD is a right triangle with angle ADB equals 90 degrees and angle ABD equals 60 degrees.
△ABD　直角三角形　∠ADB＝90°

⑤ The proportion of the length of the 3 sides of a right triangle with 3 angles of 90 degrees, 60 degrees and 30 degrees will be 1 to 2 to the square root of 3.
比　長さ　$1:2:\sqrt{3}$

⑥ Because BD to AB to AD equals 1 to 2 to the
～なので

square root of 3, AD equals the square root of 3
$AD=\sqrt{3}BD=\sqrt{3}\times 3=3\sqrt{3}$
times BD equals the square root of 3 times 3 equals

3 times the square root of 3 centimeters.

⑦So, the area of triangle ABC is (1 over 2 times 6
$\left(\frac{1}{2}\times 6\times 3\sqrt{3}=\right)9\sqrt{3}$
times 3 times the square root of 3 equals) 9 times

the square root of 3 square centimeters.
平方センチメートル(cm²)

Answer ⑧9 times the square root of 3 square centimeters

78. 三平方の定理（正三角形の高さ）

問 題 ①1辺が6cmである正三角形の面積を求めなさい。

解き方 ②図1のように頂点A、B、Cを定めます。③頂点Aを通る辺BCの垂線をひき、辺BCとの交点をDとします。④△ABDは、∠ADB＝90°、∠ABD＝60°の直角三角形です。⑤3つの角が90°、60°、30°である直角三角形の3辺の長さの比は、$1:2:\sqrt{3}$になります。⑥BD：AB：AD ＝$1:2:\sqrt{3}$なので、$AD=\sqrt{3}BD=\sqrt{3}\times 3=3\sqrt{3}$(cm) ⑦よって、△ABCの面積は、$\frac{1}{2}\times 6\times 3\sqrt{3}=9\sqrt{3}$ (cm^2)

答 え ⑧$9\sqrt{3}$ cm^2

●四角「四面」の世の中であっても……

79. The Pythagorean theorem (Diagonal lines of rectangular solids)

Question

① Like in Figure 1, there is a rectangular solid ABCD-EFGH.

(Like in: ～のように　Figure: 図　there is: ～がある　rectangular solid: 直方体)

② Find the length of the diagonal line DF.

(Find: 求めなさい　length: 長さ　diagonal line: 対角線)

Figure 1(図1)

How to solve

③ First, we join D and B and think about triangle ABD. ④ From angle DAB equals 90 degrees, triangle

(First: まずは　join: 結ぶ　triangle ABD: △ABD　angle DAB equals 90 degrees: ∠DAB＝90°)

ABD is a right triangle.

(right triangle: 直角三角形)

⑤ So, BD squared equals AD squared plus AB squared equals 3 squared plus 4 squared equals 25, and BD equals positive or negative 5 and because BD is greater than 0, BD

($BD^2=AD^2+AB^2=3^2+4^2=25$　BD equals positive or negative 5: BD＝±5　because: ～なので　BD is greater than 0: BD＞0)

equals 5 centimeters. ⑥ Next, we think about triangle
センチメートル(cm)　次に
DFB. ⑦ From angle DBF equals 90 degrees, triangle DFB is a right triangle. ⑧ So, DF squared equals DB squared plus BF squared equals 5 squared plus 3 squared equals 34, DF equals positive or negative the square root of 34, and because DF is greater than 0, DF equals the square root of 34 centimeters.

Answer ⑨ Square root of 34 centimeters

79. 三平方の定理（直方体の対角線）

問　題 ①図1のような直方体 ABCD－EFGH があります。②対角線 DF の長さを求めなさい。

解き方 ③まずは、B と D を結び、△ABD について考えます。④∠DAB＝90° より、△ABD は直角三角形です。⑤よって、$BD^2=AD^2+AB^2=3^2+4^2=25$、BD＝±5で、BD＞0より、BD＝5cm です。⑥次に、△DFB について考えます。⑦∠DBF＝90° より、△DFB は直角三角形です。⑧よって、$DF^2=DB^2+BF^2=5^2+3^2=34$、DF＝$\pm\sqrt{34}$ で、DF＞0より、DF＝$\sqrt{34}$ cm です。

答　え ⑨$\sqrt{34}$ cm

How to remember the circular constant

円周率の覚え方

①When we find the area of a circle and the length of the circumference, we use the circular constant. ②The circular constant is calculated as 3.14. ③In Japan, the circular constant is remembered by the following pun. ④*Mi hitotsu yo hitotsu iku ni muimi iwaku naku mi fumi ya yomu niroyo sanzan yami ni naku.* ⑤Like in Japan, is the circular constant remembered by a pun in the English speaking countries? ⑥Unlike in Japanese, in English, an alphabet and a word are not assigned to each numeral. ⑦In English, the numerals are equivalent to the same as the number of letters of each word. ⑧For example, there is a sentence like this, "How I wish I could calculate pi!" ⑨The number of letters of each word of this sentence is "How(3) I(1) wish(4) I(1) could(5) calculate(9) pi!(2)" ⑩When we read only the numerals in order, it becomes 3.141592. ⑪In other words, it means the circular constant up to 7 digits.

---

①円周の長さや円の面積を求めるときには、円周率を使います。②円周率は3.14として計算します。③日本では、円周率を次のような語呂合わせで覚えます。④身(3) 1(1)つ 世(4) 1(1)つ 生(5) く(9) に(2) 無(6) 意(5) 味(3) い(5) わ(8) く(9) な(7) く(9) 身(3) ふ(2) み(3) や(8) 読(4) む(6) 似(2) ろ(6) よ(4) さん(3) ざん(3) 闇(83) に(2) な(7) く(9)。⑤英語圏の国でも、日本のように円周率を語呂合わせで覚えるのでしょうか。⑥英語では、日本語のように数字１つ１つにアルファベットや単語を割りあてることはしません。⑦英語では、各単語の文字数に数字を対応させます。⑧たとえば、How I wish I could calculate pi!（どれほどパイを計算したいことか！）という文があります。⑨この文の各単語の文字数は、How(3) I(1) wish(4) I(1) could(5) calculate(9) pi!(2) です。⑩数字だけを続けて読むと、3.141592となります。⑪つまり、７桁までの円周率になっているというわけです。

Chapter 5
Use of data

第5章　データの活用

● 「度数分布表」で、何が見える化できるの？

80. Frequency distribution table・Cumulative frequency

① As in the following table, when a certain group of
～のように　次の　表　…するとき　ある～　集団
data is separated into several classes and the
資料　～に分けられる　いくつかの　階級
frequency of each class is made into a table, it is
度数　それぞれの　～につくられる
called a frequency distribution table.② By expressing
～と呼ばれる　度数分布表　表わすこと
them in a frequency distribution table, it is easier to
より簡単な
understand deviations in the number of people and
わかる　かたより　数　人
the ratio of each class, and it is easier to see the
割合　見る
state of distribution and tendencies.
ようす　傾向

Class (height) (cm) 階級（身長）	Frequency (people) 度数（人）	Relative frequency 相対度数	Cumulative frequency (people) 累積度数（人）
over　under 150.0～155.0	3	0.12	3
155.0～160.0	5	0.20	8
160.0～165.0	9	0.36	17
165.0～170.0	6	0.24	23
170.0～175.0	2	0.08	25
Total	25	1.00	

③ The value of the frequency of a certain class
値
divided by the total frequency is called the relative
割られた　合計の　相対度数
frequency.④ The relative frequency of the over 160
～以上

centimeters under 165 centimeters class is 9 divided
～未満 9÷25=0.36
by 25 equals *0 point 36. ⑤The value of the total frequency from the smallest class to a certain class
最小の
is called the cumulative frequency.
累積度数
⑥The cumulative frequency of the over 160 centimeters under 165 centimeters class is 3 plus 5
3+5+9=17
plus 9 equals 17.

*0 point 36：zero point three six と読む。

80. 度数分布表・累積度数

①次の表のように、ある集団の資料をいくつかの階級に分け、各階級の度数を表にしたものを度数分布表（どすうぶんぷひょう）といいます。②度数分布表に表わすことで、人数のかたよりや階級ごとの割合がわかりやすくなり、分布のようすや傾向が見えやすくなります。③ある階級の度数を、度数の合計で割った値を相対（そうたい）度数といいます。④160.0cm 以上165.0cm 未満の階級の相対度数は9÷25＝0.36です。⑤最小の階級から、ある階級までの度数を合計した値を累積（るいせき）度数といいます。⑥160.0cm 以上165.0cm 未満の階級の累積度数は3＋5＋9＝17です。

●びんの王冠とコイン、表が出る「確率」が高いのはどっち?

81. Statistical probability

統計的な

① The following table is a summary of the number of times a bottle cap came up *heads when tossed many times and its ratio.

次の 表 まとめ 数 回 びん キャップ 出た 表 ～したとき 投げられた 何度も 割合

Number of times tossed (times)	100	200	500	1000	2000
Number of times it came up heads (times)	42	77	198	403	801
Ratio of coming up heads	0.420	0.385	0.396	0.403	0.401

② From this table, we see that as the number of times the bottle cap is tossed increases, the ratio of coming up heads approaches 0 point 40. ③ When conducting experiments and observations in which the results depend on chance, the expressed number of how much some event is expected to happen is called the probability of that event happening. ④ *P* being probability means that when the same experiment or observation is repeated

わかる ～するにつれて 増える ～に近づく 0.40 行なう 実験 観察 結果 ～しだいである 偶然 表わされた どれくらい～か ことがら 期待される 起こる ～と呼ばれる 確率 意味する 同じ くり返される

*heads : 表裏の「表」の意。表裏を表わすときは、しばしば複数形で表現する。ちなみに「裏」は tails。

many times, the ratio of that event happening will approach p without limit. ⑤It is thought that for things like coins in which heads and tails have the same form, the probability of coming up heads will approach 0 point 5.

without: ～なしで　limit: 限界　It is thought that: …と考えられる　like: ～のような　coins: コイン　tails: 裏　form: 形　0 point 5: 0.5

*P being probability :「p が確率であること」の意で、この文の主語になっている。being は動名詞で、P は being の意味上の主語である。

びんの王冠やペットボトルのキャップなど、表と裏で形が違うものは、表と裏の出方にかたよりが出るのが一般的である。

81．統計的確率

①次の表は、びんの王冠（キャップ）を何回か投げたとき、表が出た回数と割合をまとめたものです。

投げた回数(回)	100	200	500	1000	2000
表が出た回数(回)	42	77	198	403	801
表が出た割合	0.420	0.385	0.396	0.403	0.401

②この表から、びんの王冠を投げる回数を増やしていくと、表が出る割合は、0.40に近づいていくことがわかります。③結果が偶然に左右される実験や観察を行なうとき、あることがらが起こると期待される程度を数で表わしたものを、そのことがらの起こる確率といいます。④確率が p であるということは、同じ実験や観察を多数回くり返すとき、そのことがらが起こる割合が限りなく p に近づくという意味をもっています。⑤コインのように、表と裏の形が同じものは、表が出る確率は0.5に近づくと考えられます。

●「さいころ」は投げられた?!

82. Probabilities (dice)

Question

① 2 dice, 1 large and 1 small, are rolled.
さいころ 大きい 小さい ふられる

② Find the probability that the sum of the appearing dots is a multiple of 3.
求めなさい 確率 和 出現する 目 倍数

How to solve

③ Out of n ways of all cases, when an event A happens in a ways, the probability that A happens is expressed as p equals a over n.
～のうち n 通り 場合 ～するとき ことがら 起こる a 通り 表わされる ～として $p=\frac{a}{n}$

④ When 2 dice are rolled, the total ways of appearing dots are (6 times 6 equals) 36.
全部の (6×6=)36

⑤ Among these, there are 12 ways that the sum of the appearing dots is a multiple of 3, (1, 2), (1, 5), (2, 1), (2, 4), (3, 3), (3, 6), (4, 2), (4, 5), (5, 1), (5, 4), (6, 3) and (6, 6).
～の中で ～がある

⑥ So, the wanted probability is (12 over 36 equals) 1 over 3.
求められている　$\left(\frac{12}{36}=\right)\frac{1}{3}$

Answer ⑦1 over 3

82. 確率（さいころ）

問　題　①大小2つのさいころをふります。②出る目の数の和が3の倍数になる確率を求めなさい。

解き方　③すべての場合の n 通りのうち、あることがらAが起こる場合が a 通りであるとき、Aの起こる確率 p は、$p=\frac{a}{n}$ で表わされます。④2つのさいころをふったとき、目の出方は全部で、6×6＝36(通り)です。⑤このうち、出る目の数の和が3の倍数になるのは、(1, 2)、(1, 5)、(2, 1)、(2, 4)、(3, 3)、(3, 6)、(4, 2)、(4, 5)、(5, 1)、(5, 4)、(6, 3)、(6, 6)の12通りです。⑥よって、求める確率は、$\frac{12}{36}=\frac{1}{3}$ です。

答　え ⑦$\frac{1}{3}$

●「くじ運」がいい人って、本当にいるの？

83. Probabilities (Drawing lots)

Question

① We draw lots in which 2 out of 6 are hits, 2 times one after another.
引く くじ ～のうち 当たり ～回 次々に

② Find the probability of drawing misses both times.
求めなさい 確率 引くこと はずれ 両方の

How to solve

③ We write tree diagrams like Figure 1, thinking 2 hits as A and B, and 4 misses as C, D, E and F.
樹形図 図 考えて ～として

Figure 1（図1）

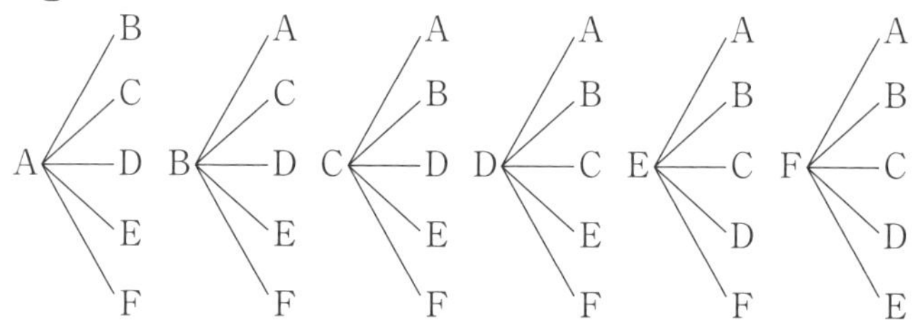

④ From Figure 1, there are 30 ways in all to draw lots.
～がある 30通り 全部で

⑤ Among these, there are 12 ways in which both
これらのうち

times are misses, (C, D), (C, E), (C, F), (D, C), (D, E), (D, F), (E, C), (E, D), (E, F), (F, C), (F, D) and (F, E).

⑥Therefore, the wanted probability is (12 over 30 equals) 2 over 5.
よって　求められている　$\left(\frac{12}{30}=\right)\frac{2}{5}$

Answer ⑦2 over 5

83. 確率（くじ引き）

問　題　①６本中２本当たりが入っているくじを２回連続で引きます。②２回ともはずれを引く確率を求めなさい。

解き方　③２本の当たりを A、B、４本のはずれを C、D、E、F として、図１のように樹形図を書きます。
④図１より、くじの引き方は、全部で30通りです。⑤このうち、２回ともはずれとなるのは、(C, D)、(C, E)、(C, F)、(D, C)、(D, E)、(D, F)、(E, C)、(E, D)、(E, F)、(F, C)、(F, D)、(F, E)の12通りです。⑥よって、求める確率は、$\frac{12}{30}=\frac{2}{5}$ です。

答　え　⑦$\frac{2}{5}$

●起こる確率と「起こらない確率」は、表裏一体なんだ

84. Probability of not happening

Question

① Find the probability that at least 1 coin comes up
求めなさい　確率　少なくとも　硬貨　出る
tails when 3 coins are tossed in order.
裏　～するとき　投げられる　順番に

How to solve

Figure 1（図1）

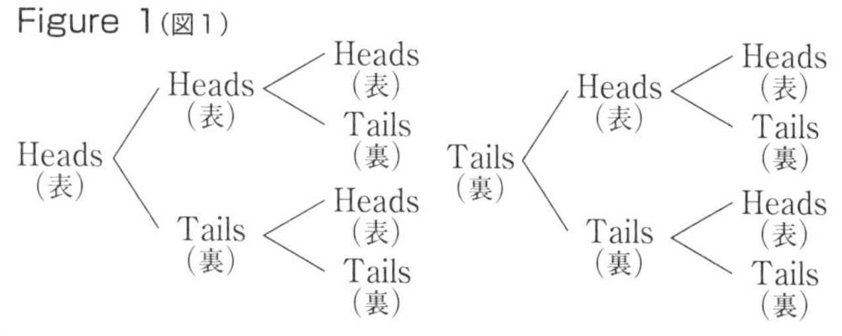

② From Figure1, there are all 8 possible ways when 3
図　～がある　起こりうる　場合
coins are tossed. ③ From these, there is 1 way when all
3 coins come up heads. ④ Therefore, the probability
表　よって
that all 3 coins come up heads is 1 over 8. ⑤ The
$\frac{1}{8}$
probability of event A not happening can be found in
ことがら　起こらない
the following way.

⑥ (probability of A not happening) = 1 − (probability
(A の起こらない確率) = 1 − (A の起こる確率)

of A happening) ⑦The probability that at least 1 coin comes up tails is the same as the probability that 'all 3 coins coming up heads' does not happen. ⑧Therefore, the probability that at least 1 coin comes up tails is 1 minus 1 over 8 equals 7 over 8.

same as：～と同じ
1 minus 1 over 8 equals 7 over 8：$1-\frac{1}{8}=\frac{7}{8}$

Answer ⑨7 over 8

7 over 8：$\frac{7}{8}$

84. 起こらない確率

問　題　①3枚の硬貨を順番に投げるとき、少なくとも1枚は裏が出る確率を求めましょう。

解き方　②3枚の硬貨を投げるとき、起こりうる場合は、図1より、全部で8通りです。③このうち、3枚とも表が出るのは1通りです。④よって、3枚とも表が出る確率は$\frac{1}{8}$です。⑤ことがらAの起こらない確率は、次のようにして求められます。⑥(Aの起こらない確率)＝1－(Aの起こる確率)⑦少なくとも1枚は裏が出る確率は、「3枚とも表が出ること」が起こらない確率と同じです。⑧よって、少なくとも1枚は裏が出る確率は、$1-\frac{1}{8}=\frac{7}{8}$ です。

答　え　⑨$\frac{7}{8}$

●「箱ひげ図」って、株のローソク足に似てない？

85. Box plots and interquartile range

①The following numbers are the scores for a 20-point test taken 10 times by a certain student arranged in ascending order.

次の　数　得点　20点満点のテスト　行なわれた　回　ある～　生徒　並べられた　小さい順に（←大きくなる順に）

8　10　12　13　13　14　15　15　17　19 (unit:point)（単位：点）

1st quartile（クォータイル）（第１四分位数）

3rd quartile（第３四分位数）

2nd quartile (median)（第２四分位数〔中央値〕）

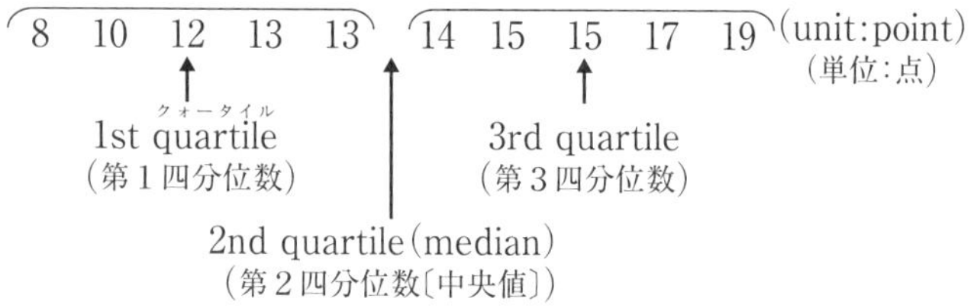

②When separated in 2 from the median, the small median of 12 points is called the 1st quartile, and the large median of 15 points is called the 3rd quartile.③And, the overall median, 13 plus 14 over 2 equals 13 point 5 (points), is called the 2nd quartile.

～したとき　分けられた　中央値　小さい　～と呼ばれる　第1四分位数　大きい　第3四分位数　全体の　$\frac{13+14}{2}=13.5$　第2四分位数

④The 1st quartile, 2nd quartile, and 3rd quartile together are called the quartile.⑤The value of the 3rd quartile minus the 1st quartile, 15 minus 12 equals 3 (points), is called the interquartile

まとめて　四分位数　値　15－12＝3　四分位範囲

range.⑥A figure expressing the minimum value, maximum value, and quartile in the following way is called a box plot.

図　表わしている　最小値　最大値　方法　箱ひげ図（←箱の平面図）

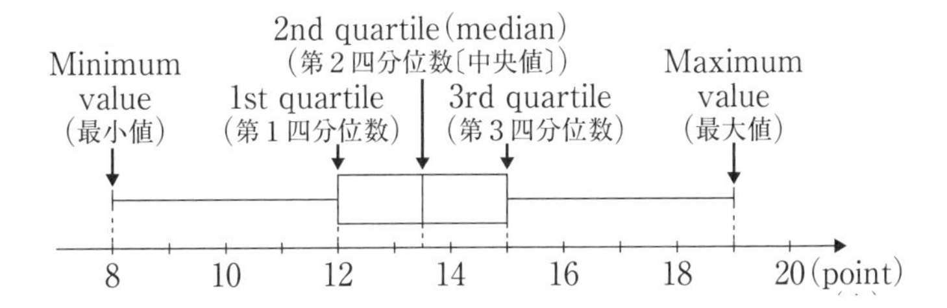

85. 箱ひげ図と四分位範囲

①次の数は、ある生徒が受けた20点満点のテスト10回分の得点を小さい順に並べたものです。②中央値を境にして2つに分けたとき、小さいほうの中央値12点を第1四分位数（しぶんいすう）、大きいほうの中央値15点を第3四分位数といいます。
③また、全体の中央値 $\frac{13+14}{2}=13.5$（点）を第2四分位数といいます。④第1四分位数、第2四分位数、第3四分位数をまとめて四分位数といいます。
⑤第3四分位数から第1四分位数を引いた値15－12＝3（点）を四分位範囲といいます。
⑥最小値、最大値、四分位数を次のように表した図を箱ひげ図といいます。

●統計の「調査」って、やっぱりあてにならない?!

86. Sample surveys and complete surveys

Question

① Which of the following 3 surveys is a complete survey and which is a sample survey?
following 次の～ surveys 調査 complete survey 全数調査 sample survey 標本調査

② (1) An audience rating survey of TV programs
audience rating 視聴率 programs 番組

③ (2) A health examination at school
health examination 健康診断

④ (3) A Cabinet approval rating survey
Cabinet 内閣 approval rating 支持率

How to solve

⑤ To examine about all of a group that is the target of a survey is called a complete survey. ⑥ To pick and examine about a part of a group is called a sample survey. ⑦ (1) In a TV audience rating survey, because it takes effort and cost to survey the programs watched on all TVs, the whole is guessed by examining a part selected at random. ⑧ So, it is a sample survey. ⑨ (2) A health examination at school is a complete survey because all students take it.
examine 調べる group 集団 target 対象 is called 呼ばれる pick 取り上げる part 部分 because ～なので takes かかる effort 手間 cost コスト survey 調査する watched 観られた whole 全体 is guessed 推測される examining 調べること selected 選ばれた at random 無作為に take 受ける

⑩(3) As for the approval rating of the Cabinet,
～についていえば
because it is difficult to examine about all voters, a
有権者
part selected at random is examined. ⑪So, it is a
調べられる
sample survey.

Answer ⑫(1) a sample survey (2) a complete survey (3) a sample survey

86. 標本調査と全数調査

問題 ①次の(1)～(3)の調査は、全数調査、標本調査のどちらですか。
②(1)テレビ番組の視聴率調査
③(2)学校での健康診断
④(3)内閣の支持率調査

解き方 ⑤調査の対象となっている集団のすべてについて調べることを全数調査といいます。⑥集団の一部分を取り出して調べることを標本調査といいます。⑦(1)テレビの視聴率調査は、すべてのテレビについて視聴番組を調査することは手間もコストもかかるので、無作為に抽出した一部分を調べて全体を推測しています。⑧よって、標本調査です。⑨(2)学校での健康診断は、生徒全員が受けるので、全数調査です。⑩(3)内閣の支持率については、有権者全員について調べることは難しいので、無作為に抽出した一部分を調べています。⑪よって、標本調査です。

答え ⑫(1)標本調査 (2)全数調査 (3)標本調査

●「近似値」って、ざっくりした値のことなんだ

87. Approximate values and significant figures

①When the volume of the juice
～するとき　体積　ジュース
in a bottle was measured with a
ビン　はかられた　～で
measuring cup graduated every
計量カップ　目盛りをつけられた　～ごとに
10 milliliters like the figure on
ミリリットル　～のような　図
the right, it was approximately 120 milliliters.
右　およそ

200mL
150mL
100mL
50mL

②As in the value of 120 milliliters gotten from
～ように　得られた
measuring with a measuring cup, the value that is not
the true value but is close to it is called the
真の　～に近い　呼ばれる
approximate value. ③The difference between the
近似値（←およその値）　差　～と…の間の
approximate value and the true value is called the
error. ④Of the numbers that express the approximate
誤差　～のうち　数　表わす
value, the numbers that are reliable are called the
信頼できる
significant figures. ⑤As there were graduations every
有効数字（←意味のある数字）～なので　～があった　目盛り
10 milliliters for the volume of 120 milliliters of juice
measured in the measuring cup, the significant figures
are the hundreds place, 1, and the tens place, 2. ⑥When
百の位　十の位

making clear *how far the significant figures go, it is
～にする　はっきりした　どれくらいまで～か
expressed in a form such as the following.⑦ (integer
形　～のような　次に述べるもの
part is a single digit number)×(the power of
(整数部分が１けたの数)×(10 の累乗)
10)⑧ For the volume of the juice measured in the measuring cup, it is expressed 1 point 2 times 10
1.2×10^2mL
milliliters squared.

*how far the significant figures go：「有効数字がどこまで行くのか」の意。有効数字の範囲を距離で比喩的に表現している。

87．近似値と有効数字

①右の図のような10mL ずつ目盛りが入っている計量カップで、ビンに入っているジュースの体積をはかったところ、およそ120mL でした。②計量カップで測定して得られた120mL という値のように、真の値ではないがそれに近い値を近似値（きんじち）といいます。③近似値から真の値を引いた差を誤差（ごさ）といいます。④近似値を表わす数字のうち、信頼できる数字を有効数字（ゆうこうすうじ）といいます。⑤計量カップではかったジュースの体積120mL では、目盛りが10mL ずつなので、百の位（くらい）の1と十の位の2が有効数字です。⑥どこまで有効数字なのかをはっきりさせるときは、次のような形で表わします。⑦(整数部分が1けたの数)×(10の累乗（るいじょう）)⑧計量カップではかったジュースの体積の場合、1.2×10^2mL と表わします。

The origin of commonly used letters in mathematics

数学でよく使われる文字の由来

① In mathematics, when numbers and quantities are expressed with letters, established letters are usually used. ② For example, n is usually used as the letter for expressing natural numbers and whole numbers, and it is the initial letter of natural number. ③ The ℓ , h, and r, used in figures are the initial letters of length, height, radius respectively. ④ It is said that the S for expressing area is the initial letter of Surface area or Sum, Summation. ⑤ *Note that it is common in English to express area with the initial letter of Area, A. ⑥ V is the initial letter of Volume. ⑦ *Expressing Pi, π comes from περίμετρος and is the Greek word for circumference. ⑧ Expressing the origin in a coordinate plane, O is taken from Origin. ⑨ *Used when expressing points in a figure, P is the initial letter of Point, and, expressing the midpoint of a line segment, M is the initial letter of Midpoint. ⑩ Expressing functions, f from $f(x)$ is the initial letter of function. ⑪ Commonly used as a variable, t comes from time.

---

①数学では数や量を文字で表わすときに、慣習的に決まった文字を使うことが多いです。②たとえば、自然数や整数を表わす文字として n を使うことが多いですが、これは natural number の頭(かしら)文字です。③図形で使われる長さ ℓ、高さ h、円の半径 r はそれぞれ length、height、radius の頭文字です。④面積を表わす S は、Surface area、または Sum、Summation の頭文字であるといわれています。⑤なお、英語では Area の頭文字 A で面積を表わすのが一般的です。⑥体積 V は Volume の頭文字です。⑦円周率を表わす π は、ギリシャ語で円周という意味のπεριμετροςに由来します。⑧座標平面で原点を表わす O は Origin からとられています。⑨図形上の点を表わすときに使われる P は Point、線分の中点を表す M は Midpoint の頭文字です。⑩関数を表わす f(x) の f は function の頭文字です。⑪変数としてよく使われる t は time に由来します。

* Note that ～：「なお、～」と説明を付け加える表現。
* Expressing ～ / Used ～：「～するとき／～されるとき」という意の分詞構文。

Chapter 6
High school mathematics for beginners

第6章　高校数学入門

● 「不等号の向き」って、気まぐれに変わるわけじゃないんだ

88. How to solve an inequality

Question

①Let's solve the following linear inequalities.
解く 次の 1次不等式

②(1) $5x-7>x+17$ (5x minus 7 is greater than x plus 17)

③(2) $2x+7\geqq 9x+28$ (2x plus 7 is greater than or equal to 9x plus 28)

How to solve

④(1) First, in the same way as a linear equation, when calculated after moving the term with the letter to the left side and the term with only numbers to the right side, it becomes 4x is greater than 24. ⑤Next, divide both sides by 4, the coefficient of x. ⑥Because 4 is a positive number, the direction of the inequality sign does not change, and x is greater than 6. ⑦(2) *When rearranged, it becomes negative 7x is greater than or equal to 21. ⑧Divide both sides by negative 7, the coefficient of x. ⑨Because negative 7 is a negative number, the

まず ～と同じように 1次方程式 ～するとき 計算される 動かすこと 項 ～のある 文字 左辺 ～だけ 数字 右辺 ～になる $4x>24$ 次に 割る 両方の 係数 ～なので 正の 向き 不等号 変わる $x>6$ 整理される $-7x\geqq 21$ -7 負の

*When rearranged：When it is rearranged の it is が省略された形。

direction of the inequality sign is reversed, and x is
逆にされる
less than or equal to negative 3.
$x \leqq -3$

Answer

⑩ (1) $x > 6$ (x is greater than 6)

(2) $x \leqq -3$ (x is less than or equal to negative 3)

88. 不等式の解き方

問　題　①次の1次不等式を解きましょう。
②(1) $5x - 7 > x + 17$　③(2) $2x + 7 \geqq 9x + 28$

解き方　④(1)まずは1次方程式と同じようにして、文字のある項を左辺に、数字だけの項を右辺に移項して計算すると、$4x > 24$となります。⑤次にxの係数4で両辺を割ります。⑥4は正の数なので、不等号の向きは変わらず、$x > 6$です。
⑦(2)移項して整理すると$-7x \geqq 21$となります。⑧xの係数-7で両辺を割ります。⑨-7は負の数なので、不等号の向きは逆になり、$x \leqq -3$です。

答　え　⑩(1) $x > 6$　(2) $x \leqq -3$

●全員集合～！したら、全体「集合」になるんだ

89. Sets

① Sets in mathematics show groups of elements that
集合 数学 示す 集まり 要素
cannot be separated any further. ② For example, the
分けられる これ以上 たとえば
elements for the set of 'single digit positive even
1けた 正の 偶数
numbers' are 2,4,6,8. ③ If we make the single digit
～すると ～を…にする
positive even numbers set A, then it is expressed as
そのとき ～と表わされる
A equals *open *brace 2, 4, 6, 8 *close brace.
A={2、4、6、8}
④ If we make the 'single digit multiples of 4' set B,
倍数
then B equals open brace 4, 8 close brace.
B={4、8}
⑤ All the elements of set B are also elements of set A.
～もまた
⑥ Then, B is called a subset of
～と呼ばれる 部分集合
A. ⑦ This kind of relationship can
この種の～ 関係
be expressed with a figure like
～で 図 ～のような
the one on the right. ⑧ This kind
右 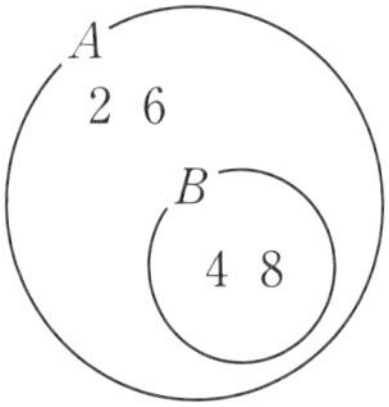
of figure is called a Venn diagram. ⑨ When
ベン図 ～するとき
considering sets, the whole set in question is called a
考える 全体の 議論されている
universal set. ⑩ For example, if we make 'all natural
全体集合 自然数

A
2 6
B
4 8

numbers' universal set U, then set A equals open brace 2, 4, 6, 8 close brace can be called a subset of universal set U.

A={2、4、6、8}

*open / close(brace):「(カッコ)を付ける／閉じる」の意。
*brace: { }を表わす単語。( )はparenthesis、[ ]はracketで表わすことが多い。

89. 集合

①数学における集合とは、それ以上細かく分けることのできない要素の集まりを指します。
②たとえば、「1けたの正の偶数」の集合における要素は、2、4、6、8です。
③1けたの正の偶数の集合をAとすると、A = {2、4、6、8}と表わします。
④「1けたの4の倍数」の集合をBとすると、B = {4、8}です。⑤集合Bの要素はすべて集合Aの要素にもなっています。⑥このとき、BをAの部分集合といいます。
⑦このような関係は、右のような図で表すことができます。
⑧このような図をベン図といいます。
⑨集合を考えるとき、考えている全体の集合を全体集合といいます。
⑩たとえば、「すべての自然数」を全体集合Uとすると、集合A = {2、4、6、8}は全体集合Uの部分集合といえます。

●「2次関数」のグラフって、彗星の軌道みたいだね

90. Quadratic functions

① When y is a function of x and y is expressed as a
～するとき　関数　～と表わされる
quadratic expression of x, y is called a quadratic
2次式　～と呼ばれる
function of x. ② In general, quadratic functions are
一般に
expressed as y equals $a\,x$ squared plus $b\,x$ plus c or y
$y=ax^2+bx+c$
equals a the *quantity x minus p close quantity
$y=a(x-p)^2+q$
squared plus q. ③ The graph for y equals a the
グラフ
quantity x minus p close quantity squared plus q,
as in Figure 1, is congruent with the graph for y
～のように　図　合同な　～と
equals $a\,x$ squared with a vertex of points (p, q), a
$y=ax^2$　～を持つ
symmetrical parabola of *vertical line x equals
対称的な　放物線　直線　$x=p$
p. ④ y equals $2x$ squared minus $4x$ minus 1 can be
$y=2x^2-4x-1$
changed into y equals 2 the quantity x minus 1
～に変えられる　$y=2(x-1)^2-3$
close quantity squared minus 3. ⑤ Therefore, the
よって
graph for y equals $2x$ squared minus $4x$ minus 1, as
in Figure 2, is congruent with the graph for y equals
$2x$ squared with a vertex of points (1, negative 3), a
頂点　-3

*quantity :「総量」の意。数式の（　）内の総量という意味で用いられている。

symmetrical parabola of vertical line x equals 1.

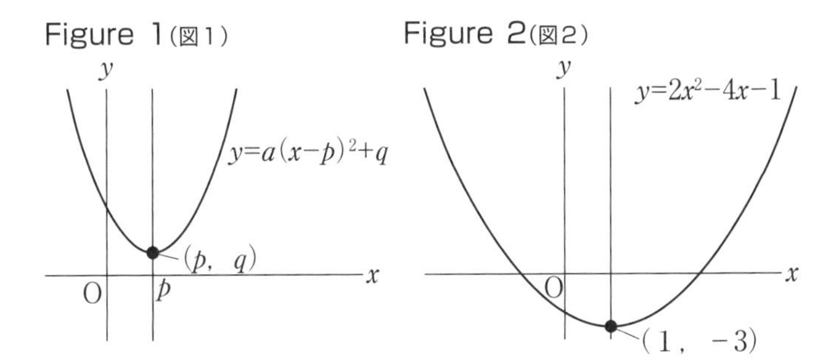

*vertical line：垂直な線。$x = p$ をグラフに表わすと垂直な線になるので、「直線」をこのように訳している。

90. 2次関数

①y が x の関数で、y が x の2次式で表わされるとき、y は x の2次関数といいます。②一般に、2次関数は $y = ax^2 + bx + c$ または $y = a(x - p)^2 + q$ と表わされます。③$y = a(x - p)^2 + q$ のグラフは、図1のように、$y = ax^2$ のグラフと合同で、頂点が点 $(p、q)$ であり、直線 $x = p$ について対称な放物線です。④$y = 2x^2 - 4x - 1$ は、$y = 2(x - 1)^2 - 3$ と変形できます。⑤よって、$y = 2x^2 - 4x - 1$ のグラフは、図2のように、$y = 2x^2$ のグラフと合同で、頂点が点 $(1、-3)$、直線 $x = 1$ について対称な放物線です。

●サイン、コサイン、タンジェント、頭文字の筆記体で覚えたね

91. Trigonometric ratios

① As for right triangle ABC, in which angle B is a
～において 直角三角形 ∠B
right angle, the ratios of the sides are decided by the
直角 比 辺 決められる
size of angle A, or A. ② If the ratios of the sides are
大きさ つまり ～すれば
known, you can find the height of a mountain, h,
知られる 求める 高さ 山
such as the one in the figure. ③ If the value of the
～のような もの（→山） 図 値
distance from the mountain, 5000 meters, and the
距離
angle at which one looks up, 24 degrees, or that BC
ある人 見上げる 24°
over AB equals 0 point 4452, is known, we know
$\frac{BC}{AB}=0.4452$
that h equals 2226 (meters) because h over 5000
h=2226 ～なので $\frac{h}{5000}=0.4452$
equals 0 point 4452.

④ From such ratios of sides, BC over AC is called
このような $\frac{BC}{AC}$ ～と呼ばれる
sine A and is expressed as sinA. ⑤ AB over AC is
サイン ～と表わされる $\frac{AB}{AC}$
called cosine A and is expressed as cosA. ⑥ BC over
コサイン $\frac{BC}{AB}$
AB is called tangent A and is expressed as
タンジェント
tanA. ⑦ The sine, cosine, and tangent are together
まとめて
called a trigonometric ratio.
（トゥリゴノメトゥリック）三角比

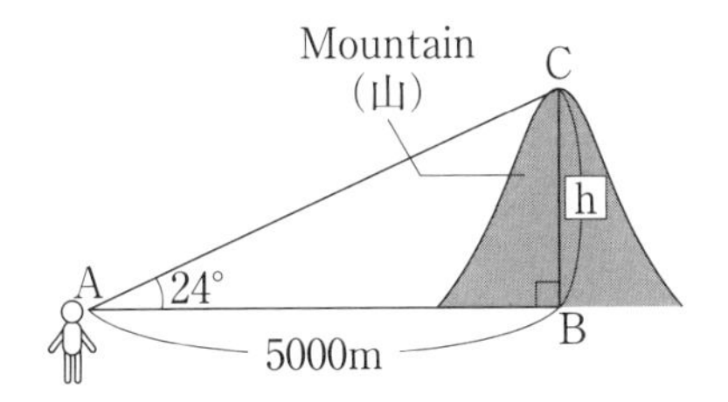

91．三角比

①∠Bが直角である直角三角形ABCにおいて、辺の比は∠Aの大きさAによって決まります。

②辺の比がわかれば、たとえば図のような山の高さhを求めることができます。

③山からの距離5000m、見上げた角度24°、$\frac{BC}{AB}=0.4452$という値がわかれば$\frac{h}{5000}=0.4452$から、h＝2226（m）とわかります。

④このような辺の比のうち、$\frac{BC}{AC}$をAのサインといい、sinAで表わします。

⑤$\frac{AB}{AC}$をAのコサインといい、cosAで表わします。

⑥$\frac{BC}{AB}$をAのタンジェントといい、tanAで表わします。

⑦サイン、コサイン、タンジェントをまとめて三角比といいます。

●スピードガンの160kmも、微分なの?

92. Differential calculus

①In Japanese, differential calculus means to ‘divide
日本語 微分（←微分の計算学） 意味する ～に分割する
into smaller parts’. ②For example, it is a way to
より小さい 部分 たとえば 方法
divide time into smaller parts and analyze the state
時間 分析する 状態
of that moment or the rate of change. ③A car
瞬間 割合 変化 クルマ
speedometer is differential calculus in itself.④This
速度計 それ自体
is because the speed displayed on a speedometer
これは…だからである 速度 表示されている
expresses the momentary state of a running car.
表わす 瞬間の 走っている
⑤Then, how is speed found? ⑥Speed equals
それでは どうやって 求められる 速度＝距離÷時間
distance divided by time.⑦If a car continues to run
～すれば ～し続ける
at the same speed, the graph becomes a straight line
同じ グラフ ～になる 直線
as in Figure 1, and the incline of that straight line
～のように 図 傾き
expresses the speed. ⑧However, in actuality, cars
しかし 実際には
do not run at a constant speed.⑨*As shown in
一定の 示される
Figure 2, the speed after, for example, 2 minutes
分
becomes equal with the incline of the tangent that
等しい ～と 接線
passes through the points of the curved line.
～を通る 点 曲線

*As shown：As it is shown の it is が省略された形。

⑩To find the incline of this tangent is to differentiate.

求める　微分する

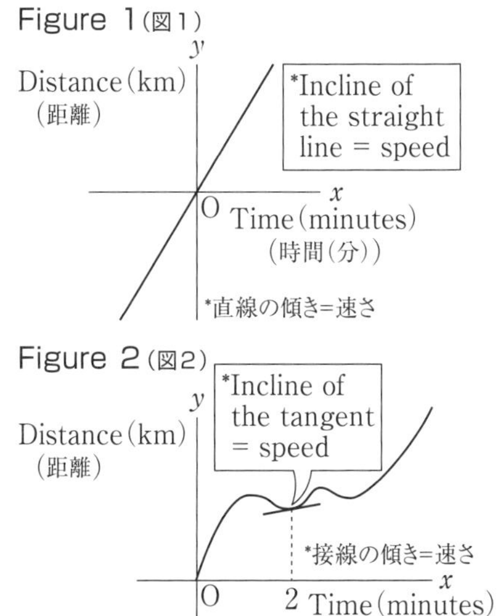

92. 微分

①微分（びぶん）とは、日本語で「微細（びさい）に分ける」という意味です。②たとえば、時間を細かく分けて、その瞬間の状態や変化の割合を分析する方法のことです。③クルマの速度計は微分そのものです。④速度計に表示されている速度は、走っているクルマの瞬間の状態を表わしているからです。⑤それでは、速度とはどのようにして求めるか。⑥速度＝距離÷時間です。⑦クルマが同じ速さで走り続ければ、図1のように、グラフは直線になり、その直線の傾きが速度を表わします。⑧しかし、実際にはクルマが一定の速度で走ることはありません。⑨図2で示すように、たとえば2分後の速度は、曲線の点を通る接線の傾きと等しくなるのです。⑩この接線の傾きを求めることが微分するということです。

●積分によって、曲線で囲まれた部分の面積が求められるんだ

93. Integral calculus

① In Japanese, integral calculus means to stack or,
日本語 積分（←積分の計算学） 意味する 積み重ねる つまり
in other words, ‘add up what was divided into
言いかえると 合計する ～なもの ～に分割された
smaller parts’. ② If the example of the car
より小さい 部分 ～するとすれば 例 クルマ
speedometer explained in the entry on differential
速度計 説明された 記載事項 微分
calculus is used, the speed, or the distance traveled
使われる 速度 走行距離（←進められた距離）
divided by the time, is differential calculus, and the
時間
distance traveled, or the speed multiplied by the
かけ算された
time, is integral calculus. ③ Integral calculus has an
inverse relationship with differential calculus.
逆の 関係 ～と
④ In other words, this means that in the case of A
場合
being differentiated and becoming B, when B is
微分されて ～になる ～するとき
integrated, it becomes A. ⑤ Integral calculus is also a
積分される ～もまた
way to find area. ⑥ As in the graph figure, when the
方法 求める 面積 ～ように グラフ 図
x-axis is time and the y-axis is speed, the shaded
x軸 濃く塗られた
area expresses the distance advanced.
表わす 進んだ
⑦ As in this graph, the distance advanced in a

certain time, even if the speed is not constant, can
ある～　たとえ～でも　一定の
be found with integral calculus.
～で

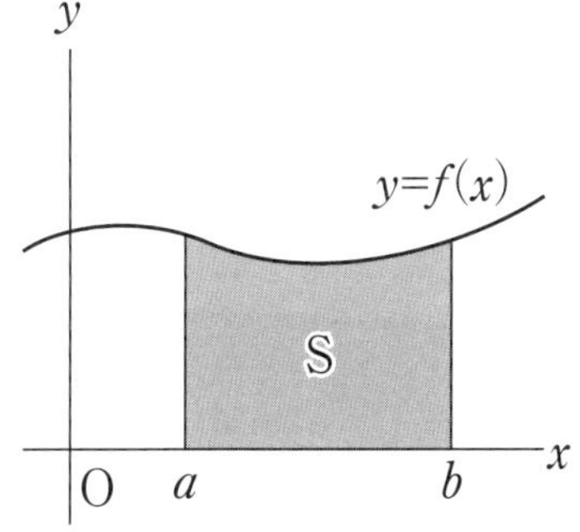

93. 積分

①積分(せきぶん)とは、日本語で積み重ねる、つまり「微細に分けたものを合計する」という意味です。②微分の項で解説したクルマの速度計の例でいえば、走行距離を時間で割った速度が微分、その速度と時間をかけ合わせた走行距離が積分ということです。③積分は微分と逆の関係になります。④つまり、Aを微分してBになるという場合、Bを積分したらAになるということです。⑤積分とは面積を求める方法でもあります。⑥図のグラフのように、x 軸を時間、y 軸を速度とすると、濃く塗られた部分の面積が、進んだ距離を表わしています。⑦このグラフのように、速度が一定でなくても、ある時間における進んだ距離は、積分で求めることができるのです。

編　者

マイプラン
1996年設立。主に学習教材の原稿執筆・編集・DTPを行なう編集プロダクション。小学校・中学校・高等学校の各教科に対応し、学校や学習塾の教材、書店売り問題集、模擬テストなど幅広い制作実績あり。学参系書籍も得意にしており、「英語対訳で読む」シリーズ（実業之日本社）や「学校では教えてくれない大切なこと」シリーズ（旺文社）などの編集も手がける。

英文執筆者

Gregory Patton（グレゴリー パットン）
1965年米国ワシントンD.C.生まれ。
コロラド大学卒業後来日、英会話学校講師を経て、現在、公立小・中学校外国語講師。

※本書は小社刊『英語対訳で読む「算数・数学」入門』を加筆・再編集したものです。

じっぴコンパクト新書　386

どう言う？　こう解く！
新版　英語対訳で読む「算数・数学」入門
"Arithmetic and Mathematics" for Beginners in Simple English (New Edition)

2021年2月10日 初版第1刷発行

編　者……………マイプラン
英文執筆者…………Gregory Patton
発行者……………岩野裕一
発行所……………株式会社実業之日本社
〒107-0062
東京都港区南青山5-4-30
CoSTUME NATIONAL Aoyama Complex 2F
電話（編集）03-6809-0452
（販売）03-6809-0495
https://www.j-n.co.jp/
印刷・製本所………大日本印刷株式会社

ISBN978-4-408-33968-9（新企画）